CULTURAL TECHNIQUES

MEANING SYSTEMS

CULTURAL TECHNIQUES

Grids, Filters, Doors, and Other
Articulations of the Real

BERNHARD SIEGERT

TRANSLATED BY
GEOFFREY WINTHROP-YOUNG

Fordham University Press : New York 2015

This publication was supported by the Internationales Kolleg für Kulturtechnikforschung und Medienphilosophie of the Bauhaus-Universität Weimar with funds from the German Federal Ministry of Education and Research. IKKM BOOKS, Volume 22. An overview of the whole series can be found at www.ikkm-weimar.de/schriften.

Fordham University Press has no responsibility for the persistence or accuracy of URLs for external or third-party Internet websites referred to in this publication and does not guarantee that any content on such websites is, or will remain, accurate or appropriate.

Fordham University Press also publishes its books in a variety of electronic formats. Some content that appears in print may not be available in electronic books.

Visit us online at www.fordhampress.com.

Library of Congress Cataloging-in-Publication Data

Siegert, Bernhard, author.
   [Essays. Selections. English]
   Cultural techniques : grids, filters, doors, and other articulations of the real / Bernhard Siegert ; translated by Geoffrey Winthrop-Young.—First edition.
        pages   cm.—(Meaning systems) (IKKM BOOKS ; Volume 22)
   Summary: "This volume designates a shift within posthumanistic media studies, that dissolves the concept of media into a network of operations, that reproduce, process and reflect the distinctions that are fundamental for a given culture, e.g. the anthropological difference, the distinctions between natural object and cultural sign, noise and information, eye and gaze"—Provided by publisher.
   Includes bibliographical references and index.
   ISBN 978-0-8232-6375-2 (hardback)—ISBN 978-0-8232-6376-9 (paper)
   1. Mass media and culture.   I. Winthrop-Young, Geoffrey, 1960– translator.   II. Title.
   P94.6S53 2015
   302.23—dc23

                                                                    2014038646

Printed in the United States of America

17   5 4 3

First edition

CONTENTS

ILLUSTRATIONS

ACKNOWLEDGMENTS

Earlier versions of chapters 1, 2, 3, 6, 7, 8, 9, and 10 and of the introduction have already appeared in print:

"Cultural Techniques, or, The End of the Intellectual Postwar in German Media Theory," trans. Geoffrey Winthrop-Young, *Theory, Culture & Society* 30 no. 6 (2013): 48–65.

"Cacography or Communication? Cultural Techniques in German Media Studies," trans. Geoffrey Winthrop-Young, *Grey Room* 29 (2007): 26–47.

"Tier essen—Gott essen—Mensch essen: Variationen des Abendmahls," in Hendrik Blumentrath et al., eds., *Techniken der Übereinkunft: Zur Medialität des Politischen* (Berlin: Kadmos, 2009), 147–68.

"*Parlêtres*: Zur kulturtechnischen Gabe und Barre der anthropologischen Differenz," in Anne von der Heiden and Joseph Vogl, eds., *Politische Zoologie* (Berlin and Zurich: Diaphanes, 2007), 23–37.

"(Nicht) Am Ort: Zum Raster als Kulturtechnik," *Thesis* 49 (2003), no. 3 (proceedings of the 9th International Bauhaus-Kolloquium, *Medium Architektur: Zur Krise der Vermittlung*, Gerd Zimmermann, ed., Weimar, 2003), 1:92–104.

"Weisse Flecken und finstre Herzen: Von der symbolischen Weltordnung zur Weltentwurfsordnung," in Daniel Gethmann and Susanne Hauser, eds., *Kulturtechnik Entwerfen: Praktiken, Konzepte und Medien in Architektur und De-sign Science* (Bielefeld: Transkript, 2009), 19–47.

"Wasserlinien: Der gekerbte und der glatte Raum als Agenten der Kon-struktion," in Jutta Voorhoeve, ed., *Welten schaffen: Zeichnen und Schreiben als Verfahren der Konstruktion* (Zürich: Diaphanes, 2011), 17–37.

"Latenz der dritten Dimension: Eine Medientheorie des Trompe-l'Oeils in der Vorgeschichte des niederländischen Stillebens," in Hans Ulrich Gumbrecht and Florian Klinger, eds., *"Latenz": Blinde Passagiere in den*

Geisteswissenschaften (Göttingen: Vandenhoeck and Ruprecht, 2011), 107–34.

"Doors: On the Materiality of the Symbolic," trans. John Durham Peters, *Grey Room* 47 (2012): 6–23.

TRANSLATOR'S NOTE

Over the years the German term *Kulturtechniken* has been rendered into English as *cultural technologies*, *cultural techniques*, and *culture technics* (with and without a hyphen). Leaving aside the differences between *Kultur* and *culture* as well as the problematic transformation of the noun *Kultur* into the adjective *cultural*, the principal quandary is the word *Technik*. Its semantic amplitude ranges from gadgets, artifacts, and infrastructures all the way to skills, routines, and procedures—it is thus wide enough to be translated as *technology*, *technique*, or *technics*. *Medientechniken*, for instance, are media technologies rather than media techniques, but *Körpertechniken* are body techniques rather than body technologies. In consultation with Bernhard Siegert I have opted in favor of *techniques*. This is not an ideal solution; in some instances my choice may well be the inferior one. However, since *Kulturtechniken* encompass drills, routines, skills, habituations, and techniques as well as tools, gadgets, artifacts, and technologies, *cultural techniques* remains the most appropriate term.[1]

CULTURAL TECHNIQUES

INTRODUCTION

*Cultural Techniques, or, the End of the Intellectual
Postwar in German Media Theory*

MEDIA THEORY IN GERMANY SINCE THE 1980S

In *The Philosophy of Symbolic Forms*, Ernst Cassirer claimed that "the critique of reason is turning into the critique of culture."[1] With the rise of so-called German media theory,[2] an alternate formula has emerged: *The critique of reason is turning into the critique of media.* Indeed, in the wake of German reunification and the subsequent countrywide reconstitution of cultural studies (*Kulturwissenschaften*), a war is waging that pits "culture" against "media." The stakes are considerable: Both combatants are striving to inherit nothing less than the throne of the transcendental that has remained vacant since the abdication of the "critique of reason." The struggle has been concealed both by a rapid succession of "turns" and by attempts to pacify combatants by introducing equalizing monikers such as "cultural media studies" (*kulturwissenschaftliche Medienforschung*). Around the turn of the millennium the war of and over German cultural studies witnessed the re-emergence of the old concept of "cultural techniques." This phrase covers a lot of what Anglophone regions like to label "German media theory." Therefore, in order to explain to the other side of the Channel and the Atlantic how this development affects so-called German media theory, it is necessary to step back and take another look at the latter.

The difficult reception of German media theory in Britain and North America was linked to the misunderstanding that it is a *theory of media,* as well as to the all-too-perceptive understanding that it never aspired to be a docile theory of media eager to join the humanities in their customary playground. What arose in the 1980s in Freiburg and has come to be associated with such names as Friedrich Kittler, Klaus Theweleit, Manfred Schneider, Norbert Bolz, Raimar

Zons, Georg-Christoph Tholen, Jochen Hörisch, Wolfgang Hagen, and Avital Ronell (and maybe also with my own) was never able to give itself an appropriate name. It definitely wasn't "media theory." One of the early candidates was "media analysis" (*Medienanalyse*), a term designed to indicate a paradigmatic replacement of both psychoanalysis and discourse analysis (thus affirming both an indebtedness to and a technologically informed distancing from Lacan and Foucault), but it just didn't work.

The "media and literature analysis"—to invoke another short-lived label— that emerged in the 1980s was not primarily concerned with the theory or history of individual media. This was already the province of individual disciplines such as film studies, television studies, computer science, radio research, and so on. Rather, its focus was literature; it strove toward histories of the mind, soul, and senses removed from the grasp of literary studies, philosophy, and psychoanalysis, and thus ready for a transfer to a different domain: media. "Media analysis as a frame of reference for other things," I read in the minutes of a 1992 meeting of the pioneers of the nameless science, convened to sketch the future shape of media research in Germany.[3] But because *media* were less a focus than a change of the frame of reference for the traditional objects of the humanities—to quote Kittler's (in)famous words, it was a matter of "expelling the spirit from the humanities"—the traditional objects of research that defined communication studies (e.g., press, film, television, radio) were never of great interest. Literature and media analysis replaced the emphasis on authors or styles with a sustained attention to inconspicuous technologies of knowledge such as index cards, writing tools, typewriters, discourse operators (including quotation marks), pedagogical media such as the blackboard, various unclassifiable media such as phonographs or stamps, instruments such as the piano, and disciplining techniques such as alphabetization. These media, symbolic operators, and drill practices were located at the base of intellectual and cultural shifts, and they primarily comprise what we now refer to as *cultural techniques*. As indicated by Hans Ulrich Gumbrecht's famous catchphrase, this changing of the humanities' frame of reference aimed to replace the hegemony of understanding, which inevitably tied meaning to a variant of subjectivity or self-presence, with "the materialities of communication"—the nonhermeneutic non-sense— as the base and abyss of meaning.[4] As a result, the focus was less *what* was represented in the media, or how and why it was represented, or why it was represented in one way rather than another. In contrast to content analysis or the semantics of representation, German media theory shifted the focus from the representation of meaning to the conditions of representation, from semantics itself to the exterior and material conditions of what constitutes semantics. Media therefore were not only an alternate frame of reference for philosophy and literature, but also an attempt to overcome French theory's fixation on discourse by turning discourse from its philosophical or archeological head onto its historical and technological feet. While Derrida's diagnosis of Rousseau's orality remained stuck in a thoroughly ahistorical phonocentrism,[5] this orality was now referred to the historico-empirical cultural technique of a maternally centered eighteenth-century oral pedagogy.[6] Derrida's *principe postale*, in turn, was no longer a metaphor for *différance*,[7] but a marked reminder that *différance* always already comes about by means of the operating principles of technical media. The exteriority of Lacan's signifier now also involved its implementation according to the different ways in which the real was technologically implemented. Last but not least, the focus on the materiality and technicality of meaning constitution prompted German media theorists to turn Michel Foucault's concept of the "historical apriori" into a "technical apriori" by referring the Foucauldian "archive" to media technologies.

This archeology of cultural systems of meaning—which some chose to vilify by affixing the ridiculous label of *media determinism* or *techno-determinism*—was (in Nietzsche's sense of the phrase) a gay science. It did not write media history, but extracted it from arcane sources (arcane, that is, from the point of view of the humanities), at a time when nobody had yet seriously addressed the concept of media. Moreover, it was archival obsession rather than passion for theory that made renegade humanities scholars focus their attention on media as the material substrate of culture. And the many literature scholars, philosophers, anthropologists, and communication experts who were suddenly forced to realize how much there was beyond the hermeneutic reading of texts when it came to understanding the medial conditions of literature and truth or the formation of humans and their souls, were much too offended by this sudden assault on their academic habitat to ask what theoretical justification lay behind this invasion.

In other words, what set German media theory on a collision course with Anglo-American media studies as well with communication studies and sociology—all of which appeared bewitched by the grand directive of social enlightenment to ponder exclusively the role of media within the public sphere—was the act of abandoning mass media and the history of communication in favor of those insignificant, unprepossessing technologies that underlie the constitution of meaning and thus elude the grasp of our usual methods of understanding. And here we come face to face with a decisive feature of this posthermeneutic turn towards the exteriority/materiality of the signifier: There is no subject area, no ontologically identifiable domain that could be called "media." Harold Innis and Marshall McLuhan already emphasized that the decision taken by communication studies, sociology, and economics to speak of media only in terms of mass media is woefully insufficient. Any approach to communication that places media exclusively within the "public sphere" (itself a fictional construct bequeathed to us by Enlightenment thought) will systematically misconstrue the abyss of nonmeaning in and from which media operate. For those eager to disentangle themselves from the grip of Critical Theory, according to which media were responsible for eroding the growth of autonomous individuality and for alienation from authentic experiences (a diagnosis preached to postwar West Germany by an opinionated conglomerate composed of the Frankfurt School, the Suhrkamp publishing house, newspapers such as *Die Zeit*, social science and philosophy departments, and bourgeois feuilletons), this abyss was referred to as "war." If the telegraph, the telephone, or the radio were analyzed as mass media at all, then it was with a view toward uncovering their military origin and exposing the negative horizon of war of mass media and their alleged public status. Hence the enthusiasm with which the early work of Paul Virilio was received in these circles (a reception that was accompanied by a lenient disregard of Virilio's pessimistically inclined anthropology).[8] Hence also the eagerness with which a materialities-based "media analysis" already early on sought out allies among those historians of science who in the 1980s abandoned the history of theory in lieu of a nonteleological history of practices and technologies enacted and performed via laboratories, instruments, and "experimental systems."[9]

"Public sphere" versus "war": This was the polemical binary under which German media theory of the 1980s assumed its distinct shape. To invoke the "public sphere" entailed ideas such as enlightened consciousness, self-determination, freedom, and so on, while to speak of "war" implied an unconscious processed by symbolic media as well as the notion that "freedom" was a kind of narcissism associated with the Lacanian mirror stage. Against the "communicative reason" as an alleged *telos* of mass media, and against the technophobe obsession with semantic depth, the partisans of the signifier unmoored from meaning and reference turned towards the history of communication engineering that had been blocked out by humanist historiography. However, the history of communication was not simply denied; it now appeared as an epoch of media rather than as a horizon of meaning. Continuing Heidegger's history of being (*Seinsgeschichte*), the history of communication was conceived of as an epoch both in the sense of a specific segment of historical time and as an *Ansichhalten* ("holding oneself back") of media.[10] The goal was to highlight the possibility of thinking media differently, that is, not only as part of the history of communication, as has been done since Karl Knies' history of the division of mental labor. Clearly, this was a departure from the usual "logocentric" narrative that starts out with the immediacy of oral communication, passes through a differentiation of scriptographic and typographic media, and then leads to the secondary orality of radio.[11]

But if media are no longer embedded in a horizon of meaning, if they no longer constitute an ontological object, how can they be approached and observed? Answer: by reconstructing the discourse networks in which the real, the imaginary, and the symbolic are stored, transmitted, and processed. Is every history of paper already a media history? Is every history of the telescope a media history? Is every history of the postal system a media history? Clearly, no. The history of paper only turns into a media history if it serves as a reference system for the analysis of bureaucratic or scientific data processing. When the chancelleries of Emperor Friedrich II of Hohenstaufen replaced parchment with paper, this act decisively changed the meaning of "power."[12] The history of the telescope, in turn, becomes a media history if it is taken as a system of reference for an analysis of seeing.[13] Finally, a history of the postal system is a media history if it serves as the system of reference for a history of communication.[14] That is to say, media do not emerge independently and outside of a specific historical practice. Yet at the same time, history is itself a system of meaning that operates across a media-technological abyss of nonmeaning that must remain hidden. The insistence on these media reference systems (designed as an attack on the reason- or mind-based humanist reference systems) was guided by a deeply antihumanist rejection of the tradition of enlightenment and the discursive rules of hermeneutic interpretation. This constitutes both a similarity and a difference between German media theory and that prominent portion of American posthumanist discourse rooted in the history of cybernetics. Within the United States, the posthuman emerged from a framework defined by the blurring of the boundaries between man and machine. However, just as U.S. postcybernetic media studies are tied to thinking about bodies and organisms, German media theory is linked to a shift in the history of meaning arising from a revolt against the hermeneutical tradition of textual interpretation and the sociological tradition of communication. Hence the cybernetically grounded American "posthuman" differed from the French "posthumanism" rooted in Heidegger, Derrida, Foucault, and Lacan, especially when taking into account their media-theoretical embeddedness. Within the framework of cybernetics, the notion of "becoming human" had as its point of departure an anthropological, stable humanity of the human that lasted until increasing feedback systems subjected the human to increasing hybridizations, in the course of which the human either turned into a servomechanism attached to machines and networks, or into a machine programmed by alien software.[15] By contrast, French (and German) posthumanism signaled that the humanities had awakened from their "anthropological slumber." As a result this type of posthumanism entailed an antihermeneutics that sought to deconstruct humanism as an occidental transcendental system of meaning production.[16] For the Germans, the means to achieve this goal were "media." The guiding question for German media theory, therefore, was not *How did we become posthuman?* but rather, *How was the human always already historically mixed with the nonhuman?*

But it was not until the new understanding of media led to the focus on cultural techniques that this variant of posthumanism was able to recognize affinities with the actor-network ideas of Bruno Latour and others. Now German observers were able to discern that something similar had happened in the early 2000s in the United States, when the advent of Critical Animal Studies and postcybernetic studies brought about a new understanding of media, as well as a reconceptualization of the posthuman as always already intertwined between human and nonhuman.

"MEDIA" AFTER THE POSTWAR ERA: CULTURAL TECHNIQUES

If the first phase of German media theory (from the early 1980s to the late 1990s) can be labeled antihermeneutic, the second phase (from the late 1990s to the present), which witnessed the conceptual transformation of media into cultural techniques, may be labeled posthermeneutic. Underneath this change, which served to relieve media and technology of the burden of having to play the bogeyman to hermeneutics and Critical Theory, there was a second rupture that only gradually came to light. The new conceptual career of cultural techniques was linked to nothing less than the end of the intellectual postwar in Germany. The technophobia of the humanities, the imperative of Habermasian "communicative reason," the incessant warnings against the manipulation of the masses by the media—all of this arose from the experiences of World War Two and came to be part and parcel of the moral duty of the German postwar intellectual. (In a talk on German postwar philosophy after Heidegger *and* Adorno at the Collège International de Philosophie in 1984, Werner Hamacher—referring to, among others, Habermas and Henrich—polemically alluded to this obligation by speaking of German "reparation payments" to Anglo-Saxon common-sense rationalism and philosophies of norms and normativity.) Given that the antihermeneutic techno-euphoria of "media analysis" and the media-materialist readings of French theory rebelled against the same set of ideas, it was no coincidence that German media theory gleefully deployed Foucauldian discourse analysis, the machinic thinking of Deleuze and Guattari, or the posthumanist Lacanian logic of the signifier against the technophobia of Critical Theory. Not surprisingly, U.S. intellectuals who had received poststructuralism as a kind of "negative New Criticism" had difficulties coming to grips with the polemical tone that permeated Kittler's writings.[17]

It was, ironicallly, the fall of the Berlin Wall and the end of the German Democratic Republic that helped redirect German postwar media theory by supplying new coordinates. Among the latter was cultural studies (*Kulturwissenschaften*), which in 1990 no longer existed in West Germany but had been practiced in the GDR, and now became one of the few Eastern heirlooms to gain acceptance in the newly united Germany. As a result, much of what perhaps should not have been referred to as *media* but was nonetheless assigned the label in order to be polemically deployed against long-standing hermeneutic aspirations and Critical Theory's yearning for a nonalienated existence, could now be designated as *cultural techniques*. The war was over—and all the index cards, quotation marks, pedagogies of reading and writing, Hindu-Arabic numerals, diagrammatic writing operators, slates, pianofortes, and so on were given a new home. This implied, first, that both on a personal and an institutional level media history and research came to abandon the shelter granted to them by literature departments. I myself left the institutional spaces of *Germanistik* in 1993 to become an assistant professor of the History and Aesthetics of Media in the re-established Institut für Kultur- und Kunstwissenschaft at Humboldt University in the former East Berlin. Second, by virtue of their promotion to the status of cultural techniques, "media" were now more than merely a "different" frame of reference for the analysis of literature, philosophy, and psychoanalysis. Third, given their new conceptual status, it now became possible to endow media with their "own" history and lay the groundwork for more systematic theoretical definitions. Fourth, with critical attention no longer focused on revealing which media technologies provide the "hard" base of the chimeras known as "spirit" (*Geist*), understanding, or the public sphere, the focus is now culture itself. Nowhere is this reorientation of German media theory more noticeable that in the changed attitude towards anthropology. During the postwar phase, anthropology was as ostracized as "man" himself, whom Kittler, for one, kept debunking as "so-called man" (*der sogenannte Mensch*). With the shift to cultural techniques, however, German media theory adopted a considerably more relaxed attitude towards a historical anthropology that relates cultural communication to technologies rather than to anthropological constants. By latching onto the old concept of cultural techniques, German media theory signals its interest in "anthropotechnics."[18]

As indicated above, this postwar turn from anti- to posthumanism appears to resemble the U.S. turn from a somewhat restricted understanding of posthumanism as a form of transhumanism (i.e., the biotechnological hybridization of human beings) to a more complex program of *posthumanities* eager to put some polemical distance between itself and old notions of the posthuman.[19] To be sure, what both turns have in common is a reluctance to interpret the "post" in posthuman in a historical sense, as something that comes "after the human." In both cases the "post" implies a sense of "always already," an onto-logical entanglement of human and nonhuman. However, the nonhuman of the German cultural techniques approach is related in the first instance to mat-ters of technique and technology, that of the American posthumanities to bi-ology and the biological. In North America the turn from the posthuman to the posthumanities is indebted to deconstruction; more to the point, it follows from the older Derrida's questioning of "the animal." In short, the German focus on the relationship between humans and machines finds its American counterpart in the questioning of the equally precarious relationship between humans and animals.[20]

But although the discussion of the man-machine-animal difference (i.e., the anthropological difference) also plays an important part in German discus-sions, and despite the links between the German understanding of cultural tech-niques and the French confluence of anthropology and technology that is now of such great importance to the American debate, critical transatlantic differ-ences remain. While the American side pursues a deconstruction of the an-thropological difference with a strong ethical focus, the Germans are more concerned with its technological or medial fabrication. From the point of view of the cultural techniques approach, anthropological difference is less the ef-fect of a stubborn anthropo-phallo-carno-centric metaphysics than the result of culture-technical and media-technological practices. The differences is espe-cially apparent in the "zoological" works of German cultural sciences that tend to be less concerned with discussions of Heidegger, Nietzsche, Agamben, and Derrida than with the media functions of animals—that is, with the way in which concrete culture techniques such as domestication and breeding, sacrificial practices, and killing methods, in connection with the emblemati-zation of certain medial virtues and capabilities of animals,[21] serve to create, shift, erode, and blur the anthropological difference.

The study of cultural techniques, however, is not aimed at removing the anthropological differences between human animal and nonhuman animal by means of subtle deconstructionist refutations of the many attempts to distin-guish between that "which calls itself human" and that "which is called ani-mal." Its goal is not to grant rights to animals, or deprive humans of certain privileges. Neither is it bent on critiquing the dogma of pure ontological dif-ference. Rather, it is concerned with decentering the distinction between hu-man and nonhuman by insisting on the radical technicity of this distinction.[22]

Human and nonhuman animals are always already recursively intertwined because the irreducible multiplicity and historicity of the anthropological is always already processed by cultural techniques and media technologies. Ahab's becoming-whale is not rooted in Herman Melville's bioethics but in the cultural technique of whale hunting. Without this technologically oriented decentering there is the danger of confusing ethics with sentimentality: The human/animal difference remains caught in a mirror stage, and the humanity that is exorcized from humans is simply transferred onto animals, which now appear as the better humans.

But what, then, were and are cultural techniques? Conceptually we may distinguish three phases.

1. Ever since antiquity the European understanding of culture implies that it is technologically constituted. The very word *culture*, derived from Latin *colere* and *cultura,* refers to the development and practical usage of means of cultivating and settling the soil with homesteads and cities.[23] As an engineering term, *Kulturtechnik*, usually translated as agricultural or rural engineering, has been around since the late nineteenth century.[24] To a certain extent the post (Cold) war turn of German media theory builds on this tradition. The corrals, pens, and enclosures that separate hunter from prey (and that in the course of coevolutionary domestication promote the anthropological difference between humans and animals), the line the plough draws across the soil, and the calendar that regulates sowing and harvesting and associated rituals, are all archaic cultural techniques of hominization, time, and space. Thus the concept of cultural techniques clearly and unequivocally repudiates the ontology of philosophical concepts. Humans *as such* do not exist independently of cultural techniques of hominization, time *as such* does not exist independently of cultural techniques of time measurement, and space *as such* does not exist independently of cultural techniques of spatial control. This does not mean that the theory of cultural techniques is anti-ontological; rather, it moves ontology into the domain of ontic operations.[25] Similar ideas relating to the production of ontological distinctions by means of ontic cultural techniques are to be found in American posthumanities, for instance, with regard to houses and the cultural techniques of dwelling.[26] This discourse, however, remains tied to the level of philosophical universals. There is no such thing as *the* house, or the house as such; there are only historically and culturally contingent cultural techniques of shielding oneself and processing the distinction between inside and outside. What (still) separates the theory of cultural techniques from those of the posthumanities, then, is that the former focuses on empirical historical objects while the latter prefers philosophical idealizations.

2. Starting in the 1970s, *Kulturtechniken* also came to refer to elementary *Kulturtechniken* or basic skills such as reading, writing, and arithmetic. Television and other information and communications technologies were added in the 1980s. What separates that particular usage of the term from its more recent application is that it still reveals a traditional middle-class understanding of culture, linking it to humanist educational imperatives. *Culture* still serves to conjure up the sphere of art, good taste, and education (*Bildung*) in a Goethean sense—in other words, it alludes to the indispensable ingredients for the formation of a "whole human." With this background in mind, the reference to television or the internet as cultural techniques aims at subjecting these new media to the sovereignty of the book—as opposed to a more pop-cultural usage that challenges the monopoly of the *alphabêtise* (Lacan) over our senses. By establishing a link with the older, technologically oriented understanding of culture, cultural techniques research breaks with a nineteenth-century middle-class tradition that conceived of culture exclusively in terms of the book reigning over all of the other arts.[27]

3. To be sure, within the new media-theoretical and culturalist context cultural techniques do refer to the so-called elementary cultural techniques, but they now also encompass the domains of *graphé* exceeding the alphanumeric code. Operative forms of writing such as calculus, cards, and catalogs, whose particular effectiveness rests on their intrinsic relationship to their material carrier (which serves to endow them with a certain degree of autonomy), are of considerable interest to those studying cultural techniques. By ascending to the status of a new media-theoretical and cultural studies paradigm, cultural techniques now also include means of time measurement, legal procedures, and the sacred. Depending on the degree to which these disciplines are affected by the "cultural turn," the concept of cultural techniques may be able to provide a systematic foundation for paleoanthropology, animal studies, the philosophy of technology, the anthropology of images, ethnology, fine arts, and the histories of science and the law.

In hindsight, the notion of cultural techniques was received—maybe all too willingly—by posthumanist cultural studies because it subverted the nonsensical war of succession between "media" and "culture" over the vacant throne of the transcendental by subjecting the two combatants to further investigation.[28] That is to say, media are scrutinized with a view toward their technicity, technology is scrutinized with a view toward its instrumental and anthropological determination, and culture is scrutinized with a view toward its boundaries, its other and its idealized notion of bourgeois *Bildung*. Against this background, and drawing upon recent discussions, we can add five further features that characterize the theoretical profile of cultural techniques.

(i) Essentially, cultural techniques are conceived of as operative chains that precede the media concepts they generate. Cultural historian Thomas Macho has remarked,

> Cultural techniques—such as writing, reading, painting, counting, making music—are always older than the concepts that are generated from them. People wrote long before they conceptualized writing or alphabets; millennia passed before pictures and statues gave rise to the concept of the image; and to this day, people may sing or make music without knowing anything about tones or musical notation systems. Counting, too, is older than the notion of numbers. To be sure, most cultures counted or performed certain mathematical operations; but they did not necessarily derive from this a concept of number.[29]

However, operations such as counting or writing always presuppose technical objects capable of performing—and to considerable extent, determining—these operations. As a historically given micronetwork of technologies and techniques, cultural techniques are the "exteriority/materiality of the signifier."[30] An abacus allows for different calculations than do ten fingers; a computer, in turn, allows for different calculations than does an abacus. When we speak of cultural techniques, therefore, we envisage a more or less complex actor network that comprises technological objects as well as the operative chains they are part of and that configure or constitute them.[31]

(ii) To speak of cultural techniques presupposes a notion of plural cultures. This is not only in deference to politically correct notions of multiculturality; it also implies a posthumanist understanding of culture that no longer posits man as the only, exclusive subject of culture. To quote a beautiful formulation by Cornelia Vismann: "If media theory were or had a grammar, that agency would find its expression in objects claiming the grammatical subject position and cultural techniques standing in for verbs."[32] Objects are tied into practices in order to produce something that within a given culture is addressed as a "person." In accordance with Philippe Descola's different "dispositives of being" (naturalism, animism, totemism, analogism), natural things, animals, images, or technological objects may also appear as persons.

(iii) In order to differentiate cultural techniques from other technologies, Thomas Macho has argued that only those techniques should be labeled cultural techniques that involve symbolic work. "Symbolic work requires specific cultural techniques, such as speaking, translating and understanding, forming and representing, calculating and measuring, writing and reading, singing and making music."[33] Indeed, the term has experienced a detrimental inflation: search engines reveal that planning, transparency, yoga, gaming, even forgetting have been promoted to cultural techniques. What separates cultural techniques from all others is their potential self-reference or "pragmatics of recursion."

> From their very beginnings, speaking can be spoken about and communication can be communicated. We can produce paintings that depict paintings or painters; films often feature other films. One can only calculate and measure with reference to calculation and measurement. And one can of course write about writing, sing about singing, and read about reading. On the other hand, it is impossible to thematize fire while making a fire, just as it is impossible to thematize field tilling while tilling a field, cooking while cooking, and hunting while hunting. We may talk about recipes or hunting practices, represent a fire in pictorial or dramatic form, or sketch a new building, but in order to do so we need to avail ourselves of the techniques of symbolic work, which is to say, we are not making a fire, hunting, cooking, or building at that very moment. Building on a phrase coming out of systems theory, we could say that cultural techniques are *second-order techniques*.[34]

It is no doubt very tempting to follow a proposal of such alluring simplicity, but unfortunately it suffers from an overly reductive notion of the symbolic in combination with a too-static distinction between *first-order* and *second-order techniques*. Granted, you cannot thematize the making of fire while making fire, but this certainly does not apply to cooking, at least not if you pay heed to Claude Lévi-Strauss's structuralist analysis. Cooking, a differentiated set of activities linked to food preparation, is both a technical procedure that brings about a transformation of the real and a symbolic act distinct from other possible acts. For instance, as part of the culinary triangle underlying the symbolic order of food preparation, the act of boiling something means to neither roast nor smoke it.[35] Hence every instance of boiling, roasting, or smoking is always already an act of communication, because it communicates to both the inside and the outside that within a certain culture certain animals are boiled, roasted, and/or smoked—like (or unlike) in other cultures, be they near or far. Because it is constituted by structural differences, cooking does indeed thematize cooking in the act of cooking.

Ploughing can be a symbolic act as well. If, as ancient sources attest, ploughs were used to draw a sacred furrow to demarcate the limits of a new city, then this constitutes an act of writing in the sense of Greek *graphé*. To plough is in this case to engage in symbolic work because the *graphein* serves to mark the distinction between inside and outside, civilization and barbarism, an inside domain in which the law prevails and one outside in which it does not. Hence doors, too, are a fundamental cultural technique, given that the operation of opening and closing them processes and renders visible the distinction between inside and outside (see chapter 10 in this volume). A door, then, is both material object and symbolic thing, a first-order as well as a second-order technique. This, precisely, is the source of its distinctive power: The door is a machine by which humans are subjected to the law of the signifier. It makes a difference, Macho writes, whether you whittle and adorn an arrow or shoot it into an animal.[36] But does this not ontologize and universalize an occidental rationality that always already separates two different types of knowledge, culture on the one hand and technology on the other? What if the arrow can be used only after it has been "decorated"? What if said "decoration" is part of the arrow's technical make-up? (See chapter 4 in this volume.)

In short, it is problematic to base an understanding of cultural techniques on static concepts of technologies and symbolic work, that is, on ontologically operating differentiations between first- and second-order techniques. Separating the two must be replaced by chains of operations and techniques: In order to situate cultural techniques *before* the grand epistemic distinction between culture and technology, sense and nonsense, code and thing, it is necessary to elaborate a *processual* rather than ontological definition of first- and second-order techniques. We need to focus on how recursive operative chains bring about a switch from first-order to second-order techniques (and back), on how nonsense generates sense, how the symbolic is filtered out of the real, or how, conversely, the symbolic is incorporated into the real, and how things/signifiers can exist because of the interchange of materials/information across the ever-emergent boundaries by which they differentiate themselves from the surrounding medium/channel.[37]

The following chapters aim to explore these processes. Macho himself alludes to the possibility of such a processual definition by speaking of *"potential self-reference."* One prime example is the art of weaving. If you adhere to the rigid distinction between first-order and second-order techniques, weaving will not qualify as a cultural technique because it does not exhibit any self-referential qualities. The term only makes sense once a piece of tapestry depicts a piece of tapestry, or a garment appears on a garment. Yet the very technique, the ongoing combination of weave and pattern, always already produces an ornamental pattern that by virtue of its technical repetition refers to itself and therefore (according to Derrida) displays sign character.[38] Following this insight, Gottfried Semper, who argued that "most of the decorative symbols used in architecture originated or were derived from the textile arts," conceived of the wall, a basic first-order architectural technique, as a second-order technique that came equipped with an originary self-reference.[39] In this way we may also distinguish

Marcel Mauss's so-called "techniques of the body" from cultural techniques,[40] that is, from the different ways in which cultures make use of bodily activities such as swimming, running, giving birth.[41] On the other hand, the recursive chains of operation that constitute cultural techniques always already contain body techniques. According to Mauss, writing, reading, and calculating, too, are techniques of the body rather than exclusively mental operations; they are the results of teaching docile bodies, which today are forced to compete with interactive navigational instruments.

(iv) Every culture begins with the introduction of distinctions: inside/outside, pure/impure, sacred/profane, female/male, human/animal, speech/absence of speech, signal/noise, and so on. The chains that make up these distinctions are recursive; that is, any given distinction may be re-entered on either side of another distinction. Thus the inside/outside distinction can be introduced on the animal side of the human/animal distinction in order to produce the distinction between domestic and wild animals. The distinction sacred/profane can be introduced on the speech side of the speech/absence of speech distinction, resulting in a split between sacred and profane languages. The constitutive force of these distinctions and recursions is the reason why the contingent culture in which we live is frequently taken to be the real, "natural" order of things. Researching cultural techniques therefore also amounts to an epistemological engagement with the medial conditions of whatever lays claim to reality. However, it is crucial to keep in mind that the distinctions in question are processed by media in the broadest sense of the word (for instance, doors process the distinction between inside/outside), which therefore cannot be restricted to one or the other side of the distinction. Rather, they assume the position of a mediating third, preceding first and second.[42] *These media are basal cultural techniques.*

In other words, the analysis of cultural techniques observes and describes techniques involved in operationalizing distinctions in the real. They generate the forms in the shape of perceptible unities of distinctions. Operating a door by closing and opening it allows us to perform, observe, encode, address, and ultimately wire the difference between inside and outside. Concrete actions serve to distinguish them from earlier nondifferentiatedness. In more general terms, all cultural techniques are based on the transition from nondistinction to distinction and back.

Yet we always have to bear in mind that the distinction between nature and culture is itself based on a contingent, culturally processed distinction. Cultural techniques precede the distinction of nature and culture. They initiate acculturation, yet their transgressive use may just as well lead to deculturalization; inevitably they partake in determining whether something belongs to the cultural domain or not. What Lévi-Strauss wrote about the art of cooking applies to all cultural techniques: "[T]he system demonstrates that the art of cooking . . . , being situated between nature and culture, has as its function to ensure their articulation one with the other."[43]

(v) Cultural techniques are not only media that sustain codes, and disseminate, internalize, and institutionalize sign systems; they also destabilize cultural codes, erase signs, and deterritorialize sounds and images. Apart from cultures of distinction, we also have cultures of de-differentiation (what once was labeled "savage" and placed in direct opposition to culture). Cultural techniques do not only colonize bodies. Tied to specific practices and chains of operation, they also serve to decolonize bodies, images, text, and music.[44] Media appear as code-generating or code-destroying interfaces between cultural orders and a real that cannot be symbolized. Resorting to a different terminology, we can refer to the nature/culture framework in terms of the real and the symbolic. By assuming the position of the third, an interface between the real and the symbolic, basal cultural techniques always already imply an unmarked space. By necessarily including the unmarked space that is excluded by the processed distinctions, cultural techniques always contain the possibility of liquidating the latter. In other words, cultural techniques always have to take account of what they exclude. For instance, upon closer scrutiny it becomes apparent that musical notational systems operate against a background of what eludes representation and symbolization—the sounds and noise of the real. Any state-of-the-art account of cultural techniques—more precisely, any account mindful of the *technological* state of the art—must be based on a historically informed understanding of electric and electronic media as part of the technical and mathematical operationalization of the real. It will therefore by necessity include what under Old European conditions had been relegated to the other side of culture: the erasure of distinctions as well as the deterritorialization and disfiguration of representations—the fall of the signifier from the height of the symbolic to the depths of the real.

The papers collected in this volume are revisions of articles and lectures written between 2001 and 2011. I will begin with a text that attempts to demonstrate how typographic, telephonic, and computer-generated media of text production may be described as cultural techniques. The three case studies that form the core of the paper focus on the specific ways in which media filter the symbolic from the real, or messages from channels full of noise. The methodological gain derived from using the cultural techniques approach is most apparent when the ontological distinction between symbols (as defined by logic) and signals (as defined by communications engineering) is replaced by the practical problem of distinguishing between them. The next four papers deal with cultural techniques related to the anthropological domain. That domain appears, first, in the context of eating. Established rituals of food intake (for which the Last Supper may serve as a paradigm for Christian-occidental culture) presuppose an already existing distinction between gods, humans, and animals, as well as between those who eat and that which is eaten. Second, the anthropological is an effect of the problematic distinction between different species of talking animals (*parlêtres*). If, as Aristotle decreed, man is an animal endowed with the gift of speech, then throughout the histories of philosophy, pedagogy, and literature this particular animal will be trailed by a host of other speaking animals (such as woodpeckers and parrots) that it has to be distinguished from—despite or because of the fact that their excluded gift of speech is always already marked as part of humanity. Third, the anthropological appears as a result of seafaring. According to Sophocles, seafaring is nothing less than a primordial cultural technology marked by a complex actor network that I will analyze using the example of the shipbuilding and navigational practices among the inhabitants of the Trobriand Islands. And fourth, the anthropological emerges as a type of subject constitution on the boundary between land and sea, Spain and America, when the bureaucratized state invented archival and notational techniques designed to make those who otherwise would have disappeared without a trace into historical darkness speak of themselves.

The next three chapters center on the importance of graphic operations as media of construction. Both "(Not) At Its Place" and "White Spots and Hearts of Darkness" focus on the decisive importance of the grid, which effectively combines an imaging process (Alberti's *velum*) with a topographical planning procedure (the colonial settlement of Latin America). It is this linking of representational and operative functions that turns the grid into a cultural technique, which can also be shown in the case of the fifteenth-century emergence of design (*disegno*) that is indebted to the link-up between the artisanal practices of Italian Renaissance artists and the rediscovered grid of longitudes and latitudes. The third text within this grouping ("Waterlines") examines how various shipbuilding techniques from the fifteenth to the nineteenth century mobilize the ontological and epistemic potential of the line.

The final two chapters are concerned with various cultural techniques of folding, opening, and closing by focusing on a specific type of operation that links media such as books, winged altars, and doors. "Figures of Self-Reference" takes as its point of departure the trompe-l'oeil technique of Dutch still life paintings in order to analyze genre-specific objects such as alcoves or tables as objects that emerged from the self-observation of the illuminated pages of Flemish books of hours. In this case we are dealing with the ontology of pictorial objects constituted by medial acts. In conclusion, the final text investigates doors as a cultural technique that processes, modifies, thwarts, and virtualizes the distinction between inside and outside. Thus the door emerges as a symbolic machine capable of weaving together diverse realities—the numinous and the profane, the imaginary and the real.

1

CACOGRAPHY OR COMMUNICATION?

Cultural Techniques of Sign-Signal Distinction

SERRES AND SIGNS

During the eighteenth century the general concept of the *sign* acted as a point of departure for the subdivision of knowledge into aesthetics (with all its internal distinctions) on the one hand and philosophical and scientific disciplines such as economy and medicine on the other. In the course of the twentieth century, however, the sign fragmented into more or less sharply separated constituted components which, in turn, became the foundational basis for a number of autonomous objects and scientific disciplines. Among the discourses arising from the decomposition of the sign, three are particularly distinct: first, the discourse of mathematical or formal logic that speaks of symbols and formal systems or languages; second, the discourse of modern linguistics and semiotics that speaks of signs, of signifiers and signifieds, synchronic systems and diachronic change;[1] and third, the discourse of communications technology that deals with signals and the physical properties of transmission conduits. While these sign disciplines are modeled on the natural sciences, the aesthetic and rhetorical aspects of the eighteenth-century semiotic discourse ended up on the side of the arts as objects of literary studies. In the early 1950s, in the wake of the new mathematical theory of communication, elements of information theory entered linguistic discourse. Saussure's distinction between *langue* and *parole* was replaced by Roman Jakobson's distinction between *code* and *message*—a terminology informed by the mathematical theory of communication and cryptology, that is, by World War sciences. In the early 1960s, however, the mathematician, philosopher, and historian of science Michel Serres proposed a simple, trifunctional model of the sign that moved the physical materiality of the channel—in other words, noise—into the center of philosophical and po-etological reflection on the sign.

In his study *The Parasite* (*Le parasite,* 1980), Serres developed the concept of the parasite into a multifaceted model that makes it possible to employ both communication theory and cultural theory to arrive at an understanding of cultural techniques. This conceptualization of the parasite is particularly interesting because it combines three different aspects. First, there is an information- or media-theoretical aspect linked to the French double mean-ing of *le parasite,* which in addition to having the same meaning as the word in English can also refer to noise or disturbance. Second, by crossing the bound-ary between human and animal, the semantics of the parasite bring into play cultural anthropology. Third, the references to agriculture and economics inherent in the term introduce the domain of cultural technology. What strikes me as revealing from the point of view of the history of theory, how-ever, is the fact that it was a reevaluation (carried out under the influence of Claude Shannon) of the Bühler-Jakobson model of communication that al-lowed Serres to sketch out a concept of cultural techniques capable of combin-ing different methods and approaches.

Serres's concept of the parasite emerged in the early 1960s when logicians were once again discussing the properties of the symbol. His initial point of departure was to replace Alfred Tarski's categorical distinction between *sym-bol,* as defined by logicians, and *signal,* as defined by information theorists, with the very *problem of distinction.* That is, Serres inquired into the conditions that enable this distinction in the first place. According to Serres, the object of in-vestigation for mathematics and logic, the symbol as *être abstrait,* is constituted by the cleansing of the "noise of all graphic form" or "cacography."[2] The condi-tions for *recognizing the abstract form* and for *rendering communication successful* are one and the same.[3] Logic, then, appears to be grounded in a culture-technical fundament that is not reflected upon.

The concept of the parasite implies a critique of occidental philosophy, in particular, a critique of those theories of the linguistic sign and economic rela-tionships that in principle never ventured beyond a bivalent logic (subject-object, sender-receiver, producer-consumer) and inevitably conceived of these relation-ships in terms of exchange. Basically, Serres enlarged this structure into a trivalent model. Let there be two stations and one channel connecting both. The parasite that attaches itself to this relation assumes the position of the third.[4] However, unlike the linguistic tradition from Locke to Searle and Habermas, Serres does not view deviation—that is, the parasite—as accidental. We do not start out with some kind of relation that is subsequently disturbed or interrupted; rather, "[t]he deviation is part of the thing itself, and perhaps it even produces

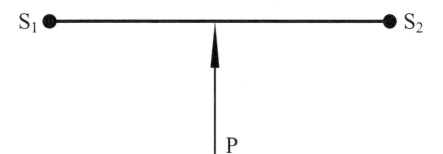

S_1 ●————————————————● S_2

P

FIGURE 1-1. Michel Serres' trivalent model of communication. Reprinted from Serres, *The Parasite*, trans. Lawrence R. Scher, 53.

the thing."[5] In other words, we do not start out with an unimpeded exchange (of thoughts, goods, or bits); rather, from the point of view of cultural anthropology, economics, information theory, and the history of writing, it is the parasite that comes first. The origin lies with the pirate rather than with the merchant, with the highwayman rather than with the highway.[6] Systems that exclude pirates, highwaymen, and idlers increase their degree of internal differentiation and are thus in a position to establish new relations. The third precedes the second: That is the beginning of media theory—of any media theory. "A third exists before the second. A third exists before the others. . . . There is always a mediate, a middle, an intermediary."[7]

In Serres's model of communication the fundamental relationship is not between sender and receiver, but between communication and noise. This corresponds to the definition of the culture-technical turn outlined above: Media are now conceptualized as code-generating interfaces between the real that cannot be symbolized and cultural orders. *"To hold a dialogue,"* Serres already wrote in 1964, *"is to suppose a third man and to seek to exclude him;* a successful communication is the exclusion of the third man."[8] Thus Serres inverts the hierarchy of the six sign functions in Jakobson's famous model (Figure 1-2).[9] It is not the poetic or the referential function that (according to the type of speech) dominates the others, but the *phatic function*, the reference to the channel. Hence in all communication each expression, appeal, and type of referencing is preceded by a reference to interruption, difference, deviation. "With this recognition the phatic function becomes the constitutive occasion for all communication, which can thus no longer be conceptualized in the absence of difference and delay, resistance, static, and noise."[10]

The phatic function—that particular function of the sign that addresses the channel—was the last of the six functions introduced by Jakobson in 1956. Its archeology once again reveals the culture-technical dimension of the

CONTEXT
(referential)

MESSAGE
(poetic)

ADDRESSER ————————————————— ADDRESSEE
(emotive)   (conative)

CONTACT
(phatic)

CODE
(metalingual)

FIGURE 1-2. Roman Jakobson's six basic functions of language. Reprinted from Jakobson, "Linguistics and Poetics," in *Style in Language*, ed. Thomas Sebeok, 353.

communication concept. It was first described in 1923 by Bronisław Malinowski, though he spoke of "phatic communion."[11] Using the communication employed during Melanesian fishing expeditions as an example, Malinowski—who in the wake of Ogden and Richards was working on a theory of meaning linked to situational contexts—developed a model of meaning that he called "speech-in-action." *Phatic communion*, however, denotes a linguistic function in the course of which words are not used to coordinate actions, and certainly not to express thoughts, but one in which a community is constituted by means of exchanging meaningless utterances. When it comes to sentences like "How do you do?" "Ah, here you are," or "Nice day today," language appears to be completely independent of the situational context. Yet there is in fact a real connection between phatic communication and situation, for in the case of this particular type of language the situation is one of an "atmosphere of sociability" involving the speakers which, however, is created by the utterances. "But this is in fact achieved by speech, and the situation in all such cases is created by the exchange of words. . . . The whole situation consists in what happens linguistically. Each utterance is an act serving the direct aim of binding hearer to speaker by a tie of some social sentiment or other."[12] The situation of phatic communion is therefore not extralinguistic, as in the case of a fishing expedition; it is the creation of the situation itself. It is a mode of language in which the situation as such appears, or in which language thematizes the "basis of relation."

There are remarkable resemblances between Malinowski's discussion of phatic communion and Serres's theory of communication, according to which communication is not the transmission of meaning by the exclusion of a third. Malinowski observes:

> The breaking of silence, the communion of words is the first act to establish links of fellowship, which is consummated only by the breaking of bread and the communion of food.[13]

Malinowski's parallel between the communion of food and the communication of words establishes an intrinsic connection between eating and speaking that is also apparent in Serres's model of the parasite.[14] For Malinowski as well as for Serres, to speak in the mode of "phatic communion" is at first merely an interruption—the interruption of silence in Malinowski's anthropological model and the interruption of background noise in Serres's information-theoretical model. Communication is the exclusion of a third, the oscillation of a system between order and chaos. Without a doubt, the link between Malinowski's phatic communion and Serres's "being of relation" (i.e., the parasite) is Jakobson's functional scheme that short-circuits the channel (in the sense of Shannon's information theory) with Malinowski's "ties of union": "The phatic function is in fact the point of contact between anthropological linguistics and the technosciences of information theory."[15]

For Serres, then, communication is not primarily information exchange, appeal, or expression, but an act that creates order by introducing distinctions; and this is precisely what turns the means of communication into cultural techniques. As stated above, every culture begins with the introduction of distinctions: inside/outside, sacred/profane, intelligible speech/barbarian gibberish or speechlessness, signal/noise. A theory of cultural techniques such as that proposed by Serres, which posits the phatic function as its point of departure, would also amount to a history and theory of interruption, disturbance, deviation. Such a history of cultural techniques may serve to create an awareness of the plenitude of a world of as-yet-undistinguished things that, as an inexhaustible reservoir of possibilities, remains the basic point of reference for every type of culture.

I will illustrate this using three examples from completely different media-historical constellations. The first example involves two elementary cultural techniques of the early modern age, the usage of zero and the typographic code; the second concerns the parasite as a message of analog channels; and the third focuses on the relationship between noise and message in digital media.

On his way to the court of the Ottoman emperor in 1555, Ogier Ghiselin de Busbecq, an ambassador in the service of Ferdinand I of Austria, came across a Latin inscription on the wall of a temple (to be precise, of a Sebasteion, a temple dedicated to a Roman emperor) in the precinct of the Haci Beiram Mosque in Angora (Ankara). Busbecq had no difficulties identifying it as copy of the famous *Index rerum gestarum*, the account of the achievements of Augustus written by the emperor himself. The heading gave away the author:

RERVM GESTARVM DIVI AVGVSTI QVIBVS ORBEM TERRARUM IMPERIO POPVLI RO-
MANI SVBIECIT ETINPENSARVM QVAS INREM PVBLICAM POPVLVMQVE ROMANVM
FECIT INCISARVM IN DVABVS AHENEIS PILIS QVAE SVNT ROMAE POSITAE EXEMPLAR
SVBIECTVM.[16]

Below is a copy of the acts of the Deified Augustus by which he placed the whole world under the sovereignty of the Roman people, and of the amounts which he expended upon the state and the Roman people, as engraved upon two bronze columns which have been set up in Rome.

The discovery of this monument of occidental cultural history, which nineteenth-century classical scholar Theodor Mommsen called the "queen of inscriptions," was by no means accidental. Throughout his journey across the Balkans and Asia Minor, Busbecq had been trying to communicate with classical antiquity. His media of communication were inscriptions and coins, his communication format was the *lectio*, in the double meaning of collecting and reading—a Judeo-Christian variant of the cultural technology that combines the cultivation of the land with the practice of reading. The biblical topos is the story of Ruth the Moabite, who plucked ears of corn left by the reapers on the field of Boaz and who was chosen to be an ancestor of King David (Ruth 2:4). Medieval monastic didactics turned Ruth the parasite into the ideal student who—to quote the prologue to the tenth-century sermons of the abbot of Morimond—by means of copying "collects the heavenly bread which is the word of God in order to satisfy the hunger of his soul."[17]

In less humble fashion, the editor responsible for the first appearance of the *Res gestae* in print spoke of the more than two hundred Greek inscriptions that Busbecq "harvested with his writing tube [*calamo exarata*]."[18] Difference and deviation have turned into cultural techniques that process residues and leftovers. Culture itself appears as a bricolage of spoils. Yet the communication with antiquity envisioned by Busbecq turns out to be a laborious venture, for the channel linking him to that antiquity is inhabited by another, more powerful parasite: the Turks. To begin with, the Turks either used antique coins as weights or melted them down to manufacture bronze vessels.[19] In addition, the Turkish transmission of biblical times and ancient Greek appeared to be, quite literally, deranged: "The Turks have no idea of chronology and dates, and make a wonderful mixture and confusion of all the epochs of history; if it occurs to them to do so, they will not scruple to declare that Job was a master of the ceremonies to King Solomon, and Alexander the Great his commander-in-chief, and they are guilty of even greater absurdities."[20] Busbecq learns in passing that history is a function of contingent cultural techniques. The Ottoman realm is just one of many possible cultures that were not realized in the Christian or European domain; in this quotation, the possible appears as the deranged, which is another name for parasitical deviation. Thus we encounter a problem that concerns the history of cultural techniques on a very basic level: namely, that history is itself an order produced by cultural techniques. As Busbecq writes, parasitical intrusions by the Turks are to blame above all for the fact that he keeps running into illegible Greek and Roman inscriptions. Such was the case with the *Monumentum Ancyranum*:

> I had it [the inscription] copied out by my people as far as it was legible. It is graven on the marble walls of a building, which was probably the ancient residence of the governor, now ruined and roofless. One half of it is upon the right as one enters, the other on the left. The upper paragraphs are almost intact; in the middle difficulties begin owing to gaps; the lower portion has been so mutilated by blows of clubs and axes as to be illegible. This is a serious loss to literature and much to be deplored by the learned.[21]

The comments written by conquerors are truly shattering.

Busbecq's copy of the *Res gestae* appeared in print for the first time in an Aurelius Victor volume edited by Andreas Schott in 1579. Humanists like Schott, who commanded the new typographic storage technology, were charged with removing fragments from stone, or from the reach of barbarian writing utensils, and, by making use of the new print medium and the system of courtly libraries, with rendering them legible enough to facilitate a new communication with antiquity undisturbed by any barbarian influence. Under these conditions, however, the real location of the letters on the interior walls of a temple surrounding the reader cannot be addressed. "Media lacunis laborare incipient"—"in the middle difficulties begin owing to gaps," Busbecq writes in his letter of 1555. "Desunt quaedam"—"a lot is missing," comments the editor in charge of the typographic reproduction of 1579. Where Busbeqc had used a locative adjective in order to speak of gaps, Schott refers to missing textual units.[22] *Desunt quaedam* doesn't point to gaps at all, but to a sign that Schott in all likelihood had invented himself:

ANNOS. VNDEVIGINTI. NATUS. EXERCITVM. PRIVATO. CONSILIO. ET. PRIVATA. IMPENSA. COMPARAVI. TERQVE. M . . . . . . .FACTIONIS. OPPRESSAM. IN. LIBERTATEM. VINDICAVI . . . . .DECRETIS. HONORIFICIS. ORDINEM. SVVM . . . .
. . . .
. . . .
. . . .
. . . .
. . . .
. . . .
. . . .*Desunt quaedam* . . . .[23]

Finally, only a few scattered words remain in an ocean of dots:

REGIS. PARTHORVM . . . .
. . . .
. . . .
. . . .
A. ME. GENTES . . . . . THORVM. ET. MEDORVM . . . .
. . . .
. . . .
. . . .
IN CONSVLATV . . . .
. . . . *Et haec quoque expunxerunt immanes Turca*. . . . .[24]

"Et haec quoque expunxerunt immanes Turcae"—"And this, too, the cruel Turks destroyed." Or, in a more literal translation, *expunged*. Was Schott aware of his double entendre, which in a bizarre way blurred the distinction between his activities and those of the barbarian commentators? It is as if the Turks, anticipating editorial interpolations, had already described the gaps as a series of dots.

In Busbecq's account there is left and right, up and down, and a center. The cultural technique of reading appears as a physical technique based on a spatial system of orientation that uses the body of the reader as its point of reference. When it comes to the edited text, however, it is no longer possible to locate the speaker. The space referred to by the commentary is linked to the gaze of a bodiless subject. To respond to the statement "A lot is missing" with the question "where?" makes no sense because the response "here" is already implicit in the comment. There is no longer any reference to a three- or two-dimensional monument. The space the commentary refers to is exactly the same space that is taken up by the commentary on paper, the space that is marked by the dots: It is the space of the text, a topological, "digitized" space.

Schott's dots uncover for all the world to see what in the case of undisturbed textual communication remains hidden: that by making use of a parasitical (supplementary) carrier, the text refers to a symbolic order based on a place-value system. There is an obvious analogy to another cultural technique, the Indo-Arabic place-value system imported by thirteenth-century Italian merchants. In our numeral system, tens, hundreds, and thousands are not explicitly written out; they are always already implicitly coded by the place that has been assigned to a digit. It is important to keep in mind that in the Indo-Arabic numeral system the spatial extension of the paper is an integral part of the numerical sign. This becomes evident in the case of zero, which marks the spatiality of the digit in the symbolic. Place-value systems are codes that take into account the media employed to store and transmit them. The channel, the parasite, is not *supplementary*, but *the ground* for the operationality of numerals. Digits are signs that can be absent from their place (as opposed to Roman numerals, which cannot be absent from their place because they have no place value). In turn, the dots introduced by Schott as signs for missing textual units are invisibly present in every letter, and only become visible when the latter is missing. Just as the invention of zero allows us to write the absence of a digit, Schott's dot is an invention that allows us to write the absence of a letter, thereby turning real gaps into a set of discrete, countable elements. The real is digitized; and the textual space is removed from barbarian cacography.

Brian Rotman has drawn attention to the close relationship between early modern algebra—as a symbolic order based on zero—and linear perspective.[25] The only position that the reading subject can assume vis-à-vis a printed text is the same that the viewing subject assumes vis-à-vis a perspectival picture. It is the position "of the Gaze, a transcendent position of vision that has discarded the body . . . and exists only as a disembodied *punctum*."[26] With this in mind, a second parallel between linear perspective and typographic textual order suggests itself. Just as Leon Battista Alberti's treatise *Della pintura* (*On Painting*) has the surface of the painting act as a window that allows us to see the objects located beyond by imposing an orthogonal grid, typographic digitization renders the monument—in Foucault's words—transparent.[27] Gazing through the printed text, we behold the true, indestructible, and complete text of the *Res gestae* in much the same way that we catch sight of the true shape of things through Alberti's window. Whereas the real still allowed for the possibility of a necessarily fragmented text, typographic coding gives rise to the notion of a necessarily complete text.[28] The third precedes the second: The typographic channel constitutes antiquity as a communication partner for humanist readers.

ANALOG MEDIA

My second example concerns a further attempt, undertaken around three hundred fifty years later, to install a communication channel between the present and Roman antiquity, Franz Kafka's famous "Pontus dream":

> Very late, dearest, and yet I shall go to bed without deserving it [Kafka writes to his fiancée Felice Bauer]. Well, I won't sleep anyway, only dream. As I did yesterday, for example, when in my dream I ran toward a bridge or some balustrading, seized two telephone receivers that happened to be lying on the parapet, put them to my ears, and kept asking for nothing but news from "Pontus"; but nothing whatever came out of the telephone except a sad, mighty, wordless song and the roar of the sea. Although well aware that it was impossible for human voices to penetrate these sounds, I didn't give in, and didn't go away.[29]

The dream represents a new version of the old invocation of the Muses.[30] It is no longer the mouth of a Homeric Muse that speaks at the origin of language, but the background noise of the telephone channel, the signal-theoretical "ground of being," as Serres would have it. No sign penetrates this noise to reach the ears of the dreamer, just an uncoded signal. That wordless song is also "the only real and reliable thing" transmitted by the phones in Kafka's *Castle*.[31] It is a message almost entirely reduced to its phatic function of referring to the channel as a nonrelating entity (i.e., as a parasite). From a technohistorical point of view, it is possible to identify this song as the voice of the telephone introduced by Philipp Reis in 1863; a reading, incidentally, supported by the context of Kafka's letter.[32] But the importance of this technohistorical reminiscence only becomes apparent once the song emanating from the receivers is deciphered as an allusion to the Siren songs of the *Odyssey*, for the latter explains the alluring and seductive quality of the song that chains the dreamer to the receivers. It is the lure of death. Kafka moves the mythic origin of language (and of culture) from the anthropological domain to that of the nonhuman, where the distinctions between language and noise, animals and humans are abolished, and which threatens—or rather, seduces—Ulysses with his own demise. The origin of language has been relocated to the realm of nonhuman signaling technology, and it is there that the dreamer hopes to hear the classical voice of Roman antiquity. For the "news from Pontus" is in fact nothing but Ovid's *Tristia*, with which the exiled poet tried to retain his *latinitas* by putting into words his

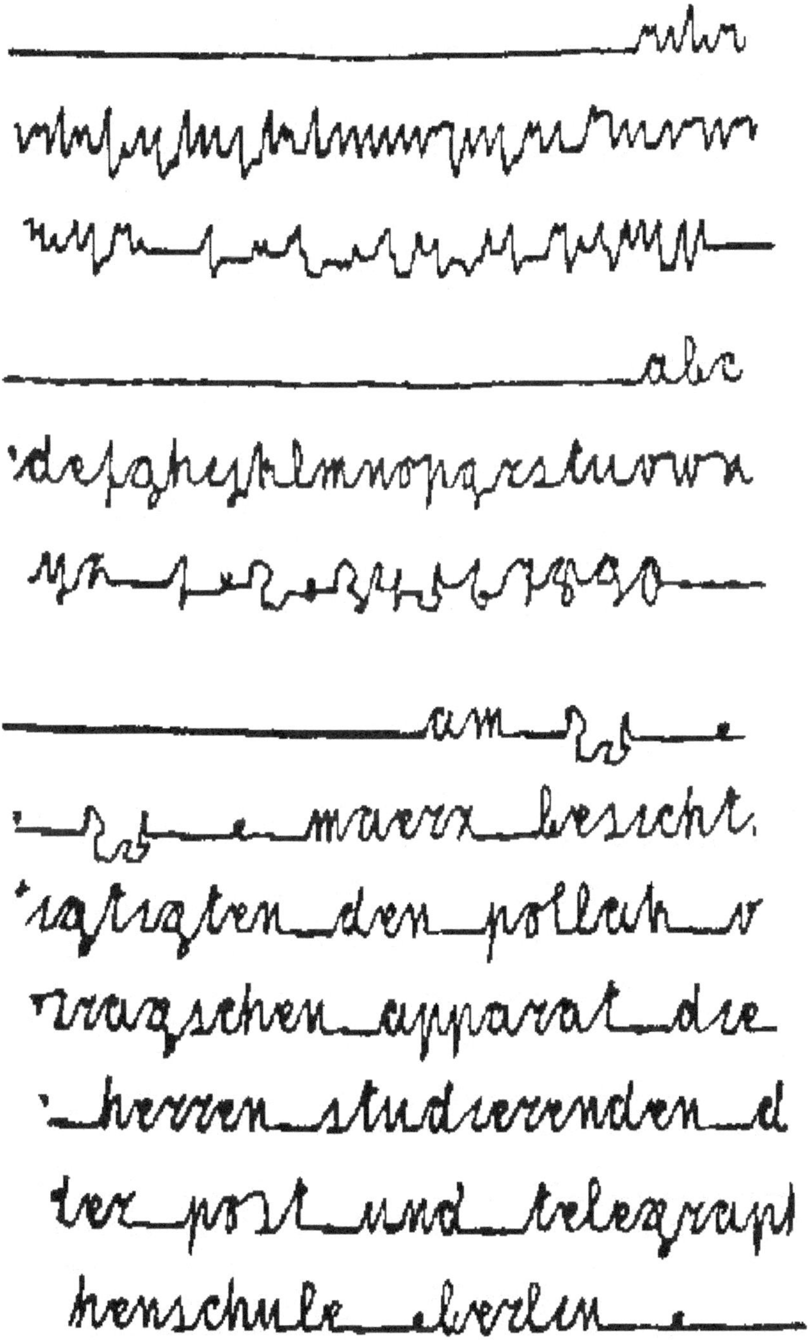

FIGURE 1-3. The Pollak / Virág telegraph: signal, character set, and sample telegram. Reprinted from Kraatz, *Maschinentelegraphen*, 102.

despair over being exiled to the Black Sea. This experience of alienation as a distance from humanity, this barbarism in the classical sense, is no longer located in the non-Latin sounds emanating from barbarian mouths; it is now based on the noise of a technical channel that human voices cannot traverse. The conceptual frame that determines the Other as well as the humanity of one's own voice has been shifted: In the age of technological media, being barbarian (*or* human) is no longer defined by the geographical and confessional boundaries of Christian Europe but by the difference between signal and noise. This, however, is a difference that alters the relationship between cultural techniques and parasites. Figure 1-3 may illustrate this: It is an ad for the telegraph developed by Pollák and Virág that was able to transmit handwritten messages, but that was only able to do so because it defined handwriting as just another cursive script or cacography.[33]

For the Pollak/Virág telegraph, handwriting is a signal much like the song of the sirens. Writing, that elementary cultural technique, emerges out of an operation that concerns the channel (the parasite) itself: It is the filtering out of signals from noise. This is, no doubt, an apocryphal example that cannot claim more than emblematic value. Yet, as my final example will clarify, the logic it illustrates becomes nothing less than systemic in the dominant cultural technique of our present: the order of digital signals.

DIGITAL MEDIA

In 1968, the Saarländische Rundfunk and Radio Bremen broadcast a radio play by Max Bense and Wolfgang Harig that presented Claude Shannon's mathematical theory of communication as an approximation to a natural language.[34] Entitled *Der Monolog der Terry Jo* (The Monologue of Terry Jo), the play referred to a girl who had been found in a boat adrift off the coast of Florida in November 1961. Though unconscious, she spoke incessantly; and the play starts out with a computer-generated text that in nine steps gradually approaches her uninterrupted flow of speech. By staging the discourse of an unconscious in such a way, the play demonstrates that in the age of signal processing, meaning is nothing but "a sufficiently complex stochastic process."[35] Shannon had demonstrated in his "Mathematical Theory of Communication" (1949) how, regardless of any grammatical deep structure or system of meaning, a natural language may be synthesized using a series of approximations, whereby the selection of a given letter depends on the probability with which it follows the preceding letter (digram structure), the two preceding letters (trigram structure), and so on.[36] *The Monologue of Terry Jo* starts out with a zero-order approximation, that is, all signs are independent of one another and equiprobable:

"fyuiömge—sevvrhvkfds—züeä-sewdmnhf—mciöwzäikmbw . . ." It then proceeds via a first-order approximation (symbols are still independent of one another but occur with the frequencies of German text) to a second-order approximation (German digram structure): "enie—sgere—dascharza—vehan—st—n—wenmen . . ."; and from there to a third-order approximation that already contains combinations of letters that look suspiciously German:

zwisch—woll möchte mit sond
ich scheid solch üb end leb gross sein und solch selb
hab hoff schluss nicht geb . . .

and so on.[37] The radio turns into a technological Muse's mouth that gives birth to language—random selections from a repertory of events with differing frequencies, from a noise whose statistical definition as an equiprobable distribution of independent signs makes it possible to interpret the channel itself as a source of information. *It* speaks.

The step leading from an analog, infinite set of signals to a finite and limitable set of selectable signals leads to the exchangeability of channel and source that is typical for the information-theoretical model of communication. Human voices may not be able to penetrate this flurry of particles, but it does allow for the synthesizing of a vocoder voice.

In a 1958 radio essay on Lewis Carroll's "Jabberwocky," Max Bense had described the inversion of the logocentric understanding of signs as a signature mark of twentieth-century media culture. While the claim of traditional metaphysical theory that "the word is the carrier of meaning" is based on the assumption that "meaning exists prior to words," Lewis Carroll was willing to maintain the "pre-existence of words—words understood as pure signals—prior to meaning."[38] As signals, words come before their meaning. Like physics, aesthetics is a science whose primary object is signals, the physical materiality of signs.

Thus a completely new understanding of the world permeating physics, logic, linguistics and aesthetics is emerging—an understanding which, briefly put, replaces

beings with frequencies
qualities with quantities
things with signs
attributes with functions
causality with statistic.[39]

"Each and every communicative relation in this world," Bense wrote in *Einführung in die informationstheoretische Ästhetik* (Introduction to information-theoretical aesthetics), "is determined as a signaling process. The world is the sum total of all signals, that is, of all signaling operations."[40] Accordingly, Bense (much like Serres, and prior to him) derives a critique of the logocentric concept of the sign. For Bense, Peirce has to be grounded in Shannon; semiotics has to be grounded in information theory. "With this signal-theoretical conception," he notes, "the sign remains a material construct."[41]

This opens up the possibility of a culture-technical approach to communication theory: The basic operation of those cultural techniques responsible for processing the distinction between nature and culture, or barbarism and civilization, is a filtering operation. If it is the goal of a communication process—be it breaking bread or breaking silence—to establish social ties by means of transcending matter and turning it into a sign, then this sign first had to be produced in the technical real. If the culture-technical operation of filtering that generates this sign from noise is in the position of a third that precedes the second and first, then Serres's work enables us to comprehend the range and impact of the current turn of cultural techniques. "We are," Serres writes in "The Origin of Language," "submerged to our neck, to our eyes, to our hair, in a furiously raging ocean. We are the voice of this hurricane, this thermal howl, and we do not even know it. It exists but it goes unperceived. The attempt to understand this blindness, this deafness, or, as is often said, this unconsciousness thus seems of value to me."[42] It is not a matter of man disappearing, but of having to define, in the wake of the epistemic ruptures brought about by first- and second-order cybernetics, noise and message relative to the unstable position of an observer. Whether something is noise or message depends on whether the observer is located on the same level as the communication system (for instance, as a receiver), or on a higher level, as an observer of the entire system. "What was once an obstacle to all messages is reversed and added to the information."[43] If exclusion and inclusion, parasite and host, are no more than states of an oscillating system or a cybernetic feedback loop, then it becomes necessary once more to inquire into those cultural techniques that, as media, process distinctions.

2

EATING ANIMALS — EATING GOD — EATING MAN

Variations on the Last Supper, or, the Cultural Techniques of Communion

THE MIXED AND THE SEPARATED

All communities are communities of the table. This is a basic axiom of Judeo-Christian-Islamic culture that appears to apply to many other cultures as well. Homer's epics leave no doubt that the rules of hospitality were instituted by Zeus in order to allow high-born members of society to engage in peaceful intercourse outside of their houses or fiefdoms without killing each other at first sight. Next to those regulating marriage, the rules determining with whom you may share your food (or not) are one of the fundamental criteria that separate cultures from one another. While sedentary, land-owning and food-producing peoples treated hospitality as an early form of "world law,"[1] the so-called service nomads (Jews, Indian Parsis, Sinti and Roma, the North African Inadan, or expatriate Chinese) used it (still use it) to clearly distinguish themselves from their surroundings.[2] But no matter what: The partaking of food for the purpose of creating a community is rarely a peaceful venture. Inevitably, something has to be killed that is then suppressed, substituted, or transfigured into a sacrifice. Not surprisingly, shared meals are marked by semiotic complexity and infinite confusion.

What do we eat when we eat together: Man, animal, or god? The Christian Eucharist, like most of antiquity's sacrificial meals, institutes a separation into theriological, anthropological, and theological domains, among which it establishes a semiotic relation. Pictorial representations—which are themselves part of this production of signs—tend to accentuate the event in different ways. The Master of the Housebook, a fifteenth-century Swabian painter and engraver, depicts in his *Last Supper* the two natures of the Passover lamb—the natural animal and its divine qualities—as co-present on the table.[3] By contrast, there is no animal left on the central part of Dieric Bouts's *Altarpiece of the Holy Sacrament* in Saint Peter's Church, Leuven (Figures 2-1 and 2-2); all we see in the bowl is a bloody residue with a few bits of bread. A vertical axis running from the ceiling's central beam through the paneling in the background into the middle fold of the tablecloth connects the objects arranged along it: the bowl that demonstrates that something that was present is now absent, the empty chalice, and a new object the Master of the Housebook had left out: the host. The painting depicts the moment in which Jesus utters the Words of Institution, *hoc est corpus meum*. One object (the animal) is missing; a new object (the host) has been added. Gleaming white, it floats over the bloody red of the bowl. The vertical axis is the axis of substitution along which two very different orders confront each other: the mixed and the separated. In the bowl containing the leftovers of the Jewish Pesach Seder, blood and bread are mixed; above it, bread and chalice are separated.

The four scenes surrounding the centerpiece are Old Testament prefigurations. To emphasize the act of substitution, the lower left wing of the triptych depicts the Passover as recorded in Exodus 12:1–28. The bowl is exactly the same, but here we see the lamb that is missing from the centerpiece.[4]

The altar is deeply invested in substitutions: the substitution of the animal by the man-god, of the man-god by the host, of the synagogue by the church, of red by white. The Leuven Confraternity of the Holy Sacrament had given

FIGURE 2-1. Dieric Bouts, *Altarpiece of the Holy Sacrament*, 1464–67. Oil on oak panel. St. Peter's Church, Leuven, Belgium. © Lukas—Art in Flanders VZW. Photo: Hugo Maertens.

FIGURE 2-2. Dieric Bouts, central panel of the *Altarpiece of the Holy Sacrament* (detail).
© Lukas—Art in Flanders VZW. Photo: Hugo Maertens.

Bouts precise instructions when it commissioned the altarpiece. In fact, the commission had been prompted by a popular cult, the *sacrament van mirakel*, that centered on a miraculous host at work in Saint Peter's Church, around which the confraternity had been founded in 1432.[5] The fourteenth century witnessed the emergence of many cults featuring bloody hosts, first in Paris and then across Europe. They originated in tales of Jewish desecration of hosts that ended in marvelous proofs of the doctrine of transubstantiation. The standard story has Jews steal a consecrated host and disfigure it using the tools of the Passion of the Christ, at which point the host begins to bleed. The perpetrators try to hide the host, but their crime is brought to light and they are executed. The Book of Hours of Mary of Burgundy depicts a host with bloody incisions (Figure 2-3). Stories of host desecrations arose against the background of the destruction of Jewish communities in Europe.[6] One such event took place in 1369–70 in Brussels. Jonathan of Eldingen, a prominent Brabantian Jew, had been charged with stealing and desecrating a host, in consequence of which the Jewish community of Leuven was destroyed on Ascension Day, 1370, and the aforementioned popular cult arose at Saint Peter's—a church, incidentally, that was located next to the destroyed synagogue. Thus a different reading of the blood in the bowl

FIGURE 2-3. Master of Mary of Burgundy, *The Miraculous Host of Dijon*, appendix to the Book of Hours of Mary of Burgundy, c. 1505. © Österreichische Nationalbibliothek, Vienna.

emerges: What humans do unto animals, they have always already done unto other humans.[7]

The community of the table is constituted by sharing, in the double meaning of the word: by dividing and apportioning the food, by inclusion and exclusion. The real animal of the Jewish Passover meal (to be seen on the lower left wing of Bout's altarpiece) is excluded by the Christian ritual in order to return as a signifier, a metaphoric sign charged with designating the "real sacrificial lamb." "Behold the Lamb of God," announces John the Baptist in the Gospel of John (John 1:29). But what happened to the lamb? Is it being digested, or did it turn itself into a sign?

According to Claude Lévi-Strauss, the axis of substitution is essential to the sacrificial meal. This appears to be a cross-cultural phenomenon, be it for Christians, the ancient Greeks, or the African Nuer, who speak of the cucumber that takes the place of the sacrificial victim as if it were an ox.[8] The sacrifice is essentially a substitute. Elias Canetti went so far as to claim that humans would never have learned to eat animals without turning into them. "The transformations which link man with the animal he eats are as strong as chains. Without transforming himself into animals he would never have learned to feed himself. . . . The flesh which is communally eaten is not what it seems to be; it stands for some other flesh and *becomes* it whilst it is being eaten."[9] Substitutions mean that the meal, insofar as it is sacrificial (in other words, a meal involving the consumption of meat), is always already a sign. The Last Supper generates cultural signs on the metonymic axis of sharing and partitioning by creating the community as a *symbolon*. In turn, the Words of Institution ensure that on the metaphoric axis of substitution, animals stand in for humans and/or god (depending on the confessional interpretations of the Eucharist). Umberto Eco emphasizes the semiotic dimension of the meal by shedding light on its underlying plan: It is said to be the result of a "design of food, not as the production of something for the individual's nourishment, but insofar as it involves the construction of contexts that have social functions and symbolic connotations, such as particular menus [or] the accessories of a meal."[10] But the axis of substitution—the sacrificial axis—does not only, as linguists would have it, construct some harmless symbolic meaning; it also inaugurates the law of the father. That is to say, community only comes about within an institutional framework; it arises by virtue of an institutionalized act of eating that is performed with a view toward honoring the name of the father: "This do in remembrance of me" (Luke 22:19). However, what is replaced in the sacrificial act of substitution is never completely replaced. As the Leuven altar shows, residues remain, an unassimilable rest indicating that the formation of community takes place against a suppressed background of murder and destruction.

The consumption of meat, then, is a ritual that creates a community as the result of (metonymic and metaphoric) acts of sharing. The sacrificial meal enacts the symbolic order of eating, thereby constituting the game of signified and signifier, transcendence and immanence, memory and pleasure.

CANNIBALISTIC BREACH OF TABOO OR DIONYSIAN CULT?

From the point of view of religious studies, the Eucharist appears to be intent on suppressing its intercultural entanglements, which become very obvious once we view it against a non-Christian background. From a Jewish perspective it constitutes a cannibalistic breach of taboo, while from a Greek point of view it smacks of the Dionysian frenzy as a manifestation of the divine. Indeed, there is reason to doubt whether Jesus really celebrated Pesach Seder with his disciples. "It is inconceivable for any Jew aware of its meaning to participate in the Eucharist without being struck dumb with horror. To Jewish culinary habits the very idea of consuming blood is abominable."[11] As a Jew, Jesus could not have spoken the words: "Take, eat, this is my body . . . Drink ye all of it, for this is my blood" (Matthew 26:26–28). Either the Jesus who presided over the Last Supper wasn't a Jew, or the most important words of the Eucharist—which do not occur in John—were invented by early Christians and retroactively attributed to him. Jesus' words, that is, the very core of the Christian ritual, would constitute a so-called community formation, courtesy of a group of inventive gentiles residing in some eastern Mediterranean port city.[12]

Bible scholars argue that Jesus' Words of Institution would have been quite acceptable to gentiles familiar with meat-consuming rituals practiced in various temples. The notion of a sacrificed god, with whom initiates merge in sacramental fashion by consuming blessed bread and sacred wine, would most likely have reminded them of religious exercises associated with certain mystery cults. The Gospel of John supports this view. John, the Greek Christian, made no secret of his anti-Judaism; he removed Jesus from his Jewish environment in order to adapt his sacrament to Greek mystery rituals. According to Jochen Hörisch, John was keen to approximate Christ and Dionysus. Nowhere is this more evident than in the displacement of the most important metaphor. John does not speak of Jesus as the "true sacrificial lamb" (that is, Christ as surpassing the Jewish Pesach sacrifice); instead he refers to him as the "true vine" (John 15:1), thus relating him to—and letting him surpass—Dionysus.[13]

"Take, eat, this is my body": The liturgical praxis adopted by Christian communities divests these words of their scandalous cannibalistic meaning by rendering them symbolic. The less temperate original utterance, however, insists that the reference to flesh and blood has to be understood literally:

> Then Jesus said unto them, Verily, verily, I say unto you, Except ye eat the flesh of the Son of man, and drink his blood, ye have no life in you. Whoso eateth my flesh, and drinketh my blood, hath eternal life; and I will raise him up at the last day. For my flesh is meat indeed, and my blood is drink indeed. He that eateth my flesh, and drinketh my blood, dwelleth in me, and I in him. (John 6:53–56)

Note that John explicitly invokes "flesh," while Mark, Matthew, and Paul resort to "body." Note also that the Greek original does not speak of "eating" and "drinking." Literally translated, John's Jesus summons his disciples to "to bite down" on his flesh and "drain" his blood (ο τρωγων μου την σαρκα και πινων μου το αιμα . . .). This scandalously theophagous cleartext reveals

Christ's Dionysian qualities. It points to central parts of the Dionysus myth, which features the Titans chewing up the flesh of the divine child, as well as to the cult of Dionysus that has the Maenads tear up raw flesh with their teeth. John's Jesus fulfills and surpasses the Eleusian mysteries rather than the promises of the Old Covenant.

The mechanics of symbolization and cultural sign production are composed of processes of abjection and transcendence. Cultural sign production is grounded in phantasms of a dismembered body and anthropophagous transgression. Disregarding Jewish dietary laws, Greek frenzy and cannibalism are designed to shape a community as a community of enemies of humanity.

AMBIVALENCE OF THE TONGUE

The various ways in which cultures creates communities (and culture itself) appear to be variations of the Last Supper. The question of how culture is rooted in acts of communal eating thus always points to the question of how the materiality of food and diet can be transcended. For Heinrich von Kleist, the Last Supper is the primal scene of the social contract agreed upon in the name of the murderous law of the father. Kleist's first play, *The Feud of the Schroffensteins*, starts out with a vow of vengeance performed as part of a Eucharist ritual, clothed in the language of the law and the social contract, to be preserved in the writing of the fathers and renewed in the covenant of the Eucharist.[14] For Kleist, obeying the law of the father means to act in utmost brutality in the name of the father, the ancestors, and the written word. In order to avoid oaths and paternal injunctions to murder, it becomes necessary to renounce one's genealogical name and give oneself a new one. In the face of the symbolic order's fall from grace grounded in the culinary ritual, the children Agnes and Ottokar represent the failed baptismal attempt to acquire new names. Kleist's play revolves around the possibility of revoking the Eucharist in order the escape the lethal sacrificial order.[15]

Perversions, contradictions, and secularizations reveal the aporias inherent in the attempt to ground culture in culinary rituals. Inevitably, the question arises of how much transcendence food, in all its base materiality, can bear. Writing to her friend Pauline Gotter, Caroline Schlegel remarked of the aesthete gourmet Carl Friedrich von Rumohr, author of the book *On the Spirit of the Art of Cooking*, "While nothing can be said against his views of cuisine, it is abominable to hear a man speak as lovingly of a crayfish as he would of the infant Jesus."[16] On the one hand, alimentary acts are embedded in the ritual of the Eucharist; on the other hand, eating is denounced as the fulfillment of our basest desires. As a result, art is left with two possible options: It can be either a cultural technique transcending matter—including flesh and blood—or a refusal of transcendence by questioning the social dimension of signs.[17]

Thus culture presents itself as the unfolding of the ambivalent relationship between language and body. The most precarious point of this relationship, the one at which it takes on all the dramatic trappings of a decisive confrontation, is the tongue. No text expresses the duplicitous cleavage of the tongue, caught between speech and food, good and bad, poison and honey, erotics and religion, more clearly than the Epistle of James:

> Even so the tongue is a little member, and boasteth great things. Behold, how great a matter a little fire kindleth! And the tongue is a fire, a world of iniquity: so is the tongue among our members, that it defileth the whole body, and setteth on fire the course of nature; and it is set on fire of hell. For every kind of beasts, and of birds, and of serpents, and of things in the sea, is tamed, and hath been tamed of mankind: But the tongue can no man tame; it is an unruly evil, full of deadly poison. Therewith bless we God, even the Father; and therewith curse we men, which are made after the similitude of God. Out of the same mouth proceedeth blessing and cursing. My brethren, these things ought not so to be. Doth a fountain send forth at the same place sweet water and bitter? Can the fig tree, my brethren, bear olive berries? (James 3:5–12)

What is the relationship between language and body? Are we—like the village community in Karen Blixen's novella *Babette's Feast*—to paraphrase this epistle and take a vow to taste nothing of the dinner from the Café Anglais? Or are we—like the officer in the same text—to merge physical enjoyment and linguistic order in an act of anagnorisis? What shall it be: articulation or the imposition of silence?

We are faced with two opposing conceptualizations of the act of eating, each acting as a point of departure for two equally divergent views of art. Eating is geared either toward acculturation or de-acculturation, formation or deformation, encoding or uncoding. Either the meal is a civilizing procedure, a semiotic disciplinary ritual (which would in particular include the history of table manners), or it constitutes an act of defiance of the semiotic regime of the social order. Subsequently, the question regarding the social significance of art is: Does art take its place on the side of transcendence or culture by turning physical acts into signs? Or is it located on the side of immanence, the nonsocial, by virtue of its insistence on nonmediation?

PHATIC COMMUNION

Not only cultural signs, but the very theory of signs as a theory of communication cannot be separated from its ethnological origin in the hegemonic model of the Last Supper. As noted in chapter 1, according to Jakobson's well-known six-function scheme, those linguistic signs that refer to their own channel have a "phatic function."[18] Phatic communication neither expresses nor references a given content; it merely ascertains the existence of the channel. It was first described in 1924 by the ethnologist Bronisław Malinowski in terms of "phatic communion": "We have here a new type of linguistic use—phatic communion I am tempted to call it . . . —a type of speech in which ties of union are created by a mere exchange of words."[19] In other words, "phatic communion" describes a deployment of language that uses words neither to coordinate actions nor to express thoughts, but that constitutes community by exchanging meaningless statements. Speakers use language to create an "atmosphere of sociability" (464). Used in such a way, language serves to foreground the situation it creates. Malinowski's explanation of phatic communion establishes a parallel with a shared meal:

> The breaking of silence, the communion of words is the first act to establish links of fellowship, which is consummated only by the breaking of bread and the communion of food. (462)

Like the breaking of a ring, the breaking of bread is invested with semiotic significance. The act of dividing turns matter into a signifier, and both halves bear witness to a signified. Breaking bread creates community in the here and now; breaking a ring creates community in absence. Breaking silence—phatic communion that refers to the communicative situation and thereby creates an atmosphere of sociability—turns the sign into a medium. Communion, the incorporation of God, which brings about a mystical *communio* of God and community, turns into a communication in which it is not the eaten God that circulates but noise. Phatic communion is therefore the interface between media technology and the sacred—a site that facilitates the translation of cultural into technical codes.

THE MEAL AS *SYMBOLON*

Unfortunately, matters are more complex. It is easy to state, as Sartre does, that "[e]very type of food is a symbol."[20] Werner Hamacher, a meticulous reader of the young Hegel, is well aware of this. Every act of eating and drinking, Hegel notes in his early writings, transcends the boundaries of the conventional sign.

The semantic reference actualized in the meal must be understood as a concrete unification.[21] Eating is not a sign that remains separate from what it signifies; rather, it is a hypermimetic sign:

> As little as the meal is a sign which remains distinct from what it signifies, so little is its ontological quality exhausted in the mere feeling of the participants that they are already unified through their common deed. The object of their common eating, that is, which as food and drink is still distinct from their community, is already conceived as this community itself. The cup of coffee or the gulp of milk is already the materially intuitable entity of the common meal.[22]

Hegel turns the paternal ritual of the Last Supper that creates the law and with it the lethal sacrificial logic of the social contract into a maternal ritual. It becomes a feast of love as imagined by the Romantics—think, for instance, of the conclusion of the Klingsohr tale in Novalis's *Heinrich von Ofterdingen*, when all imbibe the dissolved ashes of the mother who subsequently circulates among them as a unifying materialized love. The words of Jesus, "this is my body. . . . This is my blood," appear to link the principle of subjective union to material sustenance. Hegel is anxious to emphasize that this is neither an image nor an allegory "but linked to a reality."[23] If it were a mere allegory or symbol, the meal would do no more than refer; its significance would terminate in a concept. Faced with this prospect, Hegel challenges semiotic interpretations that seek to reduce the meal to a referential structure by moving the synthesis of signifier and signified into a transcendent spiritual realm beyond all physical experience.

In *Religion within the Bounds of Bare Reason* Kant implicitly denounced as a form of "fetish-making" (*Fetischmachen*) the belief that establishing a communion by means of partaking at the same table was an act "pleasing to God." Kant employs the term—which recent English translations render as *fetishism*[24]—to characterize actions "that by themselves contain nothing pleasing to God" but that are nonetheless used "as means for gaining God's direct pleasure" by pretending that they achieve a "supranatural effect through entirely natural means." Such acts, Kant continues, turn the "service of God into mere *fetishism*" (198; emphasis in the original) and the church with its attendant "*priestery*" into a place of "*fetish service*" (199; emphases in the original). The concluding "General Comment" summarizes different "kinds of delusory faith" including the ceremony "of a shared partaking at the same table." Religious censorship prevents Kant from explicitly denouncing the Last Supper ritual as "fetish-making," but the implication is clear. "But to boast that God has linked special graces with the celebration of this ceremony, and to admit among the articles of faith the proposition that this celebration, which is after all merely a church action, is yet in addition also *a means of grace*, is a delusion of religion that can do nothing other than work precisely contrary to religion's spirit" (220–21; emphasis in the original).

On the Protestant side, the primal scene of this argument is the famous Marburg Colloquy of 1529 that pitted Martin Luther against Ulrich Zwingli. For the latter, the corporeal presence of Christ in the bread and wine of the Eucharist revealed a repugnant attachment to the sensual. Zwingli therefore maintained that *hoc est corpus meum*, the Words of Institution, had to be read metaphorically and translated as "this signifies my body."[25] At this point in the discussion Luther tore off the velvet tablecloth, so that all present were able to read the words *hoc est corpus meum*, which prior to the dispute he had written with chalk on the table.[26] The Words of Institution substitute for what is usually brought to the table. Luther's dramatic gesture quotes the ceremonial revelation of the consecrated host in the Catholic ritual, when the drapes of the tabernacle are drawn aside in order to allow the populace to see the incarnated body of Christ in the shape of bread. The Colloquy's table becomes the altar; the tablecloth, in turn, mediates the presence of body and blood in the word. Something of such unquestionable presence requires no further discussion.

The meal is thus neither conventional sign nor image nor allegory: It is a symbol. But while Hegel interprets the meal as a symbol in the Greek sense, he deliberately misunderstands the Greek *symbolon*. To ensure that relationships based on mutual hospitality endure for years or even generations it is necessary that both sides agree upon signs of recognition that the stranger seeking hospitality can at a later date use to identify himself. Such signs—*symboloi* or *tessera hospitalis*—function as storage media for long-distance relationships. It could, for instance, be a ring that host and guest broke in half and shared at their first meeting; presenting their halves will enable the children and grandchildren of the guest and the host to identify themselves as descendants when they meet for the first time and resume the original friendship.[27] But in marked contrast to the rituals of antiquity, which rigorously separated the two, Hegel merges meal and symbol. The meal itself did not function as a *symbolon*; *symboloi* are storage media. Unlike a meal, they do not serve the here and now but the future. Hegel disagrees: *symbolon* and meal coincide. The symbol is no longer a storage medium but a medium of the presence of a union. The signifier is erased so that it may reappear in the real of a physical union. Hamacher notes:

> In Hegel's text, food and drink, the body and blood of Christ, on the one hand, and eating and those who eat on the other, but also, thirdly, the united

eaters and the eaters in their individual particularity, all figure as symbols, as pieces of a circle which belong together.[28]

This sounds more harmless than it is. To claim that not only food and drink but also the eaters themselves are circulating in and around the united eaters, is to conjure up the latent cannibalistic problem that serves to turn the *symbolon* into a *diabolon*. Up until now we were more or less dealing with a cultural history of eating, that is, with a cultural history that maps the relationship between the real and the symbolic. But as pointed out in the introduction to this volume, what distinguishes the focus on cultural techniques from cultural history is that the former focuses less on the ontological distinction between symbol and signal than on the technical problems underlying this distinction. According to Michel Serres, a symbol first needs to be filtered out of the cacophony of the real. What Hegel had argued in the context of the Last Supper—namely, that sharing the meal is not a conventional sign but a symbol in the real—turns out to be a problem of civilizing the sacred as well as the barbarian. As we shall see, the problem is that a meal does not result in a loving union (and thus not in any Christian community), because the symbol in the real can flip over into abjection and disgust.

POLLUTION OF THE SYMBOLIC

The question of how culture can be founded in common meals needs to be re-formulated: How can eating establish common bonds if it is caught up in the tension between phatic communion (the breaking of bread and the atmosphere of sociability) and disgust, that is, between contact and standoff? So far, we have treated dividing and sharing as relatively abstract procedures. In the context of hospitality and phatic or channel-centered communication we referred to the breaking of the bread as a symbolic act. Hegel, for one, was so interested in the metonymic transfer of the host's sacrament that he got into trouble with the Vatican by claiming that if even the crumbs of a consecrated host are sacred, the faeces of a mouse that has eaten these crumbs must be considered sacred too.

Once it is a matter of enacting this symbolic procedure in the real, cultural techniques such as cutlery, tableware, and table manners enter the picture. Media of touch and contact distribute the food across the table. In the real, the breaking of the bread—the scene of the symbol and the creation of community—is an act that contaminates the channels of community. The bread touches the food in the bowl (see Dieric Bouts's altar piece) and floats in the wine. Conduct books written between the thirteenth and fifteenth century were full of admonitions to either drain or spill the wine into which one has dunked a piece of bread. The repeated injunctions were necessary: In the Middle Ages and the early modern period, people liked to dip their bread into whatever liquid happened to be close by, and glasses, cups, and mugs were usually shared by several guests. Dipping one's bread into a wine cup meant that the neighbor was forced to drink the remaining morsels of bread. According to the Gospels Jesus partook of bread and wine and then pronounced them his "body" and "blood," but there is no mention of his saying that the body should be dipped and crumbled into the blood—in other words, there was no command to breach the boundaries between eating and drinking. The breaking of the bread, the eucharistic gesture of phatic communion, pollutes the communal channel; as a result the meaning of the gesture may suffer an inversion: Instead of creating community and sociability, it gives rise to disgust and repulsion. The sign is contaminated with the channel, the symbolic with the real. This is precisely what Hegel had described so romantically in *The Spirit of Christianity and Its Fate* and came to phrase more deliriously in his later writings. But how are we to neatly separate sign and channel? Cutlery and tableware facilitate conditions in which the sharing of food may remain a symbolic act free from all contaminations by the real (dirt, disturbance, noise). The frequently invoked transcendence of matter requires, first and foremost, the right technology.

Contact media, then, create community in the communal act of eating. This contact, however, allows for movement in two directions. The media not only distribute the communal food to participating guests, but also the participating guests to the communal food. The basic problem of the Eucharist—one that required the invention of the complex dogma of transubstantiation—is: How can we both ensure and avoid the eating of the host? In turn, the basic problem of the many ceremonial meals derived from the Eucharist is: How do you avoid eating the eaters along with the food? In the Marburg words of Zwingli: "Quod in coena se non dedit corporaliter?"—how do you avoid not physically giving yourself while eating?[29] How do you prevent the table community from eating itself? Food appears to contain an excess—namely, the metonymic presence of those consuming it. As Hegel noted, it is "linked to a reality." The imitations of the Last Supper are metonyms rather than metaphors; transubstantiation is replaced by transferring traces of the real. When all is said and done and eaten, there really is a fly in the ointment, or maybe something different and even more disgusting, be it spit or scraps of skin. We consume our food down to the skin and bones—but if we aren't careful we may consume each other. Of course this is not anthropophagy in the literal sense; human flesh is not served straight up. It arrives on the table in the shape of metonymic traces of contact with the "real" food, as constantly dissolving and redrawn physical boundaries. The perversions—or rather pollutions—of the Last Supper are as omnipresent as they are inevitable. There is the constant danger that the focus on the host and his substitutions may descend into a partial cannibalism in which everybody eats everybody. Jean-Jacques Rousseau's *Confessions* offer a famous piece of literary evidence for the attempt to ground intimacy in cannibalistic pleasure. The scene revolves around the complex game of fetishistic longings and suppressed incestual desire directed at Rousseau's substitute mother *Maman*. Just as Maman replaces the dead mother, she in turn becomes the object of an unfulfillable desire that replaces the lost object by means of countless metonymic displacements, as Rousseau's famed orality ranges from the abjection of writing to the kissing of drapes and furniture to include the act of eating:

> Even in her presence I would sometimes commit some extravagance, which it seemed only the most violent love could have inspired. One day at the table, when she had just put a forkful of food into her mouth, I cried out that I had seen a hair on it; she rejected the mouthful and put it on the side of her plate, whereupon I seized it avidly and swallowed it.[30]

Think, also, of the famous replacement of kissing and biting in Kleist's *Penthesilea*.[31] Whether Rousseaus's public festivities, in the course of which the people became aware of themselves in public self-representation,[32] can be deduced from this primal scene is open to debate, though Derrida certainly had a point when arguing that the Swiss people, by assuming the role both of actor and audience, tend to indulge in self-incorporating autophagy. Each individual Swiss is simultaneously host and guest. Nonetheless, what the Dionysian love feast appears to celebrate as the ingestion of the other, the paternal law swiftly denounces as disgusting. What kind of medium is necessary to prevent the self-incorporation of the community (which represents the republican version of the Eucharistic incorporation of the host) from being processed in the real?

A MEDIA THEORY OF TABLEWARE

Most table manners are designed to regulate the exchange between the group and its individual members. It is imperative to rigorously detach two points of contact: that between hand and food (which is inside the pot or bowl from which all eat) and that between hand and mouth. Tableware and cutlery serve to create a singular, individual eating body that does not add itself to the food consumed by the community. In his massive study *The Civilizing Process* Norbert Elias amassed a large amount of relevant material from early modern conduct books. "Do not put back on your plate what has been in your mouth,"[33] one primer states. Others note how improper it is to use one's hand to blow one's nose. "You should not poke your teeth with your knife, as some do; it is a bad habit,"[34] is another frequent admonition, as is the reminder to spit only at the wall or under the table.

If cutlery and tableware are filters, table manners are the protocols for filtering a symbolic out of noise. Culinary parasites need to be removed from the channel; the goal is to create clear, unmistakable subjects on the one side and corresponding objects on the other. The filtering operation separates subject and objects that barbarian and Eucharist meals had mixed together. In time, this decomposition process gives rise to subject-object relations. Initially, however, this decomposition is a sovereign privilege. The sovereign is the first to be differentiated and created as a singular figure. Even when in the company of their retinue, emperors, kings, and royal princes eat alone at their table. From a sociological perspective this is done to emphasize that they have nothing in common with the rest.[35] But once we take into account the cultural techniques involved, the main reason for this privileged segregation is that insufficient isolation would result in a contamination of the body of the sovereign. The body of the king is seated apart because it is threatened by permanent contagion. The prince, therefore, is the first eating body marked by a sense of taste: by developing a distinguished way of eating he is creating, as it were, an eating self.

The royal table features a cup, a spice box, and a mustard jug; as well, there is a *nef*, a ship-shaped holder for locking away personal utensils before and after meals in order to prevent any poisoning. Precautions against poisoning are the privilege of the sovereign. The chief courtier acts as food taster; bezoar stones, adder's tongue, or cups made from narwhal tusk or rhinoceros horn indicate by discoloration or foaming whether any poison is present. The prince represents a special case within the community: He is an individual because he is one who does not eat all the others. To guarantee that he touches only what has not been touched by anyone else, he is removed from circulation. The sovereign is a sovereign because the contamination of food by exposure to others would poison him. The individualizing effect of table manners—that great theme of Norbert Elias—is a result of the republican table community ingesting the sovereign and turning the royal defense protocol into a filter system. The separation of the one from all the other eaters and the separation of his food from the food of the others turns into the separation of culinary subjects from culinary objects. From now on each of us may become an eating self, a subject that ingests others without losing itself in doing so. And this, in turn, enables the emergence of taste as a cultural technique of differentiation and distinction.

The most important demixing tool is the tablecloth. Introduced in the age of Charlemagne, it is used to wipe one's dirty fingers. Dieric Bouts's altarpiece features a second, draping tablecloth on top of the first (which appears to be freshly spread out), which functions as a common napkin. The tablecloth reveals the dialectic of table manners: It was originally used to wipe fingers and spoons. Together, tablecloth and spoon form a kind of valve that lets the soup pass from pot to mouth but keeps the body of the eater from travelling in the reverse direction. On its passage from mouth to pot the spoon has to travel by way of the tablecloth. In the bourgeois realm of table manners (in which the plate serves to interrupt the metonymic transfer of the eater back to the communal food), the tablecloth serves to completely exclude the table from all dietary channels. Good manners dictate that it must remain clean.[36] Adorned with a white cloth (the color of brides and popes), the table not only prohibits physical contact (and hence desire), it also intimates the symbolic castration that allows for the construction of a symbolic order from the circulation of the real.[37]

In other words, the Eucharist model of commensality suffers from an autoimmune reaction.[38] It threatens itself from two sides: On the one hand it may well fall short of the model of the *symbolon* as long as the latter is conceived as a meal that not only images but also *contains* the unity of commensality. This model, if pushed to the extreme, results in mutual cannibalism. On the other hand it is jeopardized by the model of the *diabolon*: The speculative movement, which the young Hegel wanted to affirm using the model of the Eucharist, may freeze in crystalline isolation, in a self-reference no dialectic can sublate. As a result of this autoimmunization of the community, the *symbolon* can only be practiced by deferral, as one not identical with itself.

EUCHARIST À LA WINDIGO

"Mortality and Mercy in Vienna," one of Thomas Pynchon's first stories, presents another variation of the Last Supper, yet in contrast to Hegel it refuses to dissolve in a speculative manner the contrast of matter and meaning, body and sign, cannibalism and community. The story denies salvational transcendence by referring back to Joseph Conrad's *Heart of Darkness*. Restricting himself to the immanence of flesh and blood, Pynchon's protagonist Cleanth Siegel turns his sacrificial cult into a critique of culture's incessant attempts to symbolize and transcend matter. "Mortality and Mercy" locates the subject at the intersection of cultural theory, anthropology, psychoanalysis, and colonialism. Siegel himself becomes a site at which Jewish and Catholic systems enter into a heretic alliance and the rituals of the Eucharist ultimately merge with Ojibwe cannibalism imported from the woods of Ontario.

Shedding all resemblance to the ritual transmitted by the gospels, Pynchon's Last Supper instead anticipates the wild parties staged by "The Whole Sick Crew" in his novels *V.*, *The Crying of Lot 49*, and *Gravity's Rainbow* with all their outlandish culinary, musical, and sexual practices. The coupling of three opposites is particularly striking: (i) in the religious sphere, that of Judaism and (Jesuitically inflected) Catholicism; (ii) in the colonial sphere, that of civilization and wilderness (Washington, D.C. and Conrad's jungle); and (iii), in the domain of anthropology and cultural theory, that of cannibalism and the Eucharist.

Ever since "Entropy,"[39] his first published story, Pynchon's parties have stood for a situation in which the communion of bread and wine and the communication of words are no longer subject to the rules of community formation but to the law of entropy. But if the information content of broken bread and sacred utterances is approximating that of noise, what is the role of the host, the representative of the law of the father? The event is bracketed by a change of roles of host and guest. At the outset of the story, Siegel, the guest, and Lupescu, the host, switch roles, a reversal Pynchon supports by invoking the different meanings of *hospes/hostis*:

> "It's all yours," he [Lupescu] said. "You are now the host. As host you are a trinity: (a) receiver of guests"—ticking them off on his fingers—"(b) an enemy and (c) an outward manifestation, for *them*, of the divine body and blood."[40]

This serves to conjure up the triple meaning of the meal in which one and the same person functions as ceremonial host, sacramental host, and enemy. Ancient Greek did not distinguish between guest and stranger—both were referred to as *xenos*, *xeinos*, or *xénä*. In Latin too it is difficult to draw a clear etymological boundary between guest and stranger.[41] The latter was originally called *hostis*, which also meant enemy; but as of the first century BCE the term only signified the (political) enemy. In Old Latin, however, it indicated members of alien tribes or nations, including those on peaceful terms with Rome, with the understanding that they were living under their own laws. The position of the arriving stranger is indeed ambivalent: On the one hand he is a sinister enemy, on the other a guest who deserves respect. The word *guest* mirrors this ambivalence. It is derived from Indo-European **ghostis*, which not only spawned Latin *hostis* but also evolved into *guest* and *ghost*. Both engage in visitations: A guest is someone who comes to haunt your house, in other words, a ghost.

The stranger who accepts the offer of hospitality refrains from using his powers against the wellbeing of the receiving party; he willingly submits to the host.

Precisely this abstention is captured in Latin *hospes*. The word is made up of *hosti-pot-s*. *Pot* is "the powerful one," "the master"; *hospes* is "he who has power over the stranger." Influenced by the Greeks, Latin culture starts to use *hospes* for both guest and host, while *hostis* is restricted to enemy. Thus the boundaries between the three are blurred.

The host, then, has a triple role. He is, first, the host in a secular, nonreligious sense: he welcomes the guests, shares his food, and listens to their stories; second, he is the enemy; and third, he is the host in the liturgical sense, the transubstantiated flesh and blood of the host (in the first sense), who lets himself be eaten to redeem humanity. The coding of the subject in the imperial tension between civilization and wilderness allows Siegel to merge all three roles into one. Conrad's *Heart of Darkness*, which is directly quoted in Pynchon's story, depicts how the colonial envoy Kurtz goes native and establishes his own private realm of horror at the periphery of the Belgian Congo, where he, the human sacrifice, awaits the priest who will be sent up the river. As depicted in Francis Ford Coppola's *Apocalypse Now* (U.S., 1979), which relocates events to Cambodia during the Vietnam War, the cult of death on the Upper Congo (or the Nung River) mirrors the barbarism of those waging colonial wars in the name of civilization. Kurtz tears off the mask of civilization to reveal the primitive savagery of the white man—and this, in turn, entails the refusal of any ritual-based symbolization. Kurtz stages an apocryphal variant of the Eucharist with himself as the victim, but while the human sacrifice takes place, there is no possibility of transcendence, no possibility of generating signs, metaphors, substitutions, or any kind of significance. This, precisely, is the meaning of his dying words, "the horror, the horror."[42] The redemptive refusal of metaphor is seized upon by Pynchon, who gives it an additional twist. It is no longer the colonial officer who reveals the heart of darkness at the center of civilization and goes mad, but Irving Loon, an Ojibwe who has been imported from the arboreal, first-order jungle to an urban, second-order jungle. Unbeknownst to his surroundings, Loon stands on the verge of an outbreak of Windigo psychosis.

Windigo psychosis is a culture-specific syndrome involving compulsive acts of cannibalism commonly associated with members of the tribes of the Algonquian language family (Chippewa, Ojibwe, and Cree) formerly inhabiting the Great Lakes region of Canada and the United States. It usually surfaced in the winter, when isolated tribes had been forced to live for months without sufficient nutrition. As a result the individual believed that he had been transformed into a wendigo, a supernatural monster that consumes human flesh. Those in the grip of Windigo psychosis increasingly viewed those around them as potential food. In Pynchon's words:

"Get the picture," he [Siegel] had told Grossmann that night, over mugs of Würtzburger. "Altered perception. Simultaneously, all over God knows how many square miles, hundreds, thousands of these Indians are looking at each other out of the corner of their eye and not seeing wives or husbands or little children at all. What they see is big fat juicy beavers. And these Indians are hungry, Grossmann. I mean, my gawd. A big mass psychosis. As far as the eye can reach"—he gestured dramatically—"beavers. Succulent, juicy, fat." (18)

Irving Loon is attending the party as a "conquest" of economics expert and sex tourist Debby Considine, who has brought him along from Ontario. When Loon addresses Debby as "my beautiful little beaver"(20), he is using a term of endearment composed of an obscene word for vulva and a meat dish. Indeed, since "beaver" denotes that part of her body with which Debby Considine achieves her conquests, the term constitutes a very special "incorporation" of a white woman as the hallucination of a meat dish. A hallucination is neither metaphor nor imagination but the reappearance in the real of a signifier that was rejected in the symbolic. If we turn to the differences between Christianity and Judaism, we recognize that from a Christian perspective this is tantamount to the difference between a first- and second-order signifier. From the Christian point of view, the relationship between Old and New Testament is one of typological prefiguration—as, for instance, in Bouts's altarpiece in Saint Peter's. Christianity (in this particular case, the Confraternity of the Holy Sacrament) interprets the Old Testament, that is, the Passover lamb, as a metaphor of the cleartext: the lamb is Jesus Christ. Siegel's fusion of Christianity and Judaism undoes this hierarchy of metaphor and truth. In the end the Jesuit voice, with its emphasis on the miraculous power of redemption, blends with the Jewish voice intonating a dirge for the dead:

He [Siegel] stood in the kitchen, alone, trying to assess things. . . . How much had Irving Loon been drinking? How much did starvation have to do with the psychosis once it got under way? And then the enormity of it hit him. Because if this hunch were true, Siegel had the power to work for these parishioners a kind of miracle, to bring them a very tangible salvation. A miracle involving a host, true, but like no holy eucharist. (21)

Siegel has the party end in a bloodbath. As he watches the Ojibwe grab a Browning army rifle, Siegel is caught between two supper rituals: His Catholic side is exhilarated "that the miracle was in his hands after all, for real"(22), while his Jewish side is lamenting the dead. Loon loads the rifle; for beaver consumption no further tableware is necessary. In this particular Last Supper, the culinary mingling of subjects and objects is nothing to be avoided; instead it is the very goal of the meal—a true republican and postcolonial Eucharist. "It was just unfortunate that Irving Loon would be the only one partaking of any body and blood, divine or otherwise" (23).

Siegel, the father, the host who throughout the evening had to play confessor to his guests, discreetly leaves the feast. The-name-of-the-father (the law), once instituted by the biblical primal scene, is now destituted by Pynchon's wendigo variant. The self-removal of the paternal law allows for the transformation of the female phallus—the "beaver"—into food. The lack is turned into edible presence and enables the party revelers to consume flesh and blood in a cannibalistic feast. After playing both attentive host (in the word's first meaning) and death-bringing enemy (second meaning), Siegel enacts the third meaning of "host": the representative of the law of the father surreptitiously takes off so that the female phallus may reenter the real body of the community, where it will appear in the shape of bloody stigmata. Reaching the street below, Siegel hears the first burst of the gun. "What the hell, stranger things had happened in Washington" (23). *Hoc est corpus meum.*

3

PARLÊTRES

The Cultural Techniques of Anthropological Difference

DESCARTES' PARROT

According to Deleuze and Guattari, the animals on display in the zoo of philosophers—the philosophical beasts, as it were—constitute a subspecies of classified animals.[1] Talking birds are kept in this menagerie for the purpose of illustrating that special "plus x" which separates humans from animals. The word *parlêtre* (a term on loan from Lacan)[2] signifies a being that subverts the anthropological difference based on language. Talking birds, however, explode the category of the philosophical animals; they may even suspend Deleuze and Guattari's entire typology.

It would be easy to describe the ups and downs of certain animals in this special zoo. For instance, it appears that in our philosophy seminars, apes have come to occupy the position once held by birds. Ever since Darwin, the burden of signifying the anthropological difference has rested on apes. This was not always in the case. In the days before the theory of evolution emerged as the anthropological paradigm, indeed already during antiquity and the Renaissance, the simian aptitude for imitation was enlisted to provide proof of what separates humans—and artists in particular—from animals.[3] Seventeenth-century philosophers, in turn, were interested less in apes than in certain bird species. The reason seems obvious. If Aristotle was right to define man as the animal that possesses *logos* (that is, language and reason),[4] then this definition must be able to withstand a comparison with other animals that also evince *logos*. The *logos* of the animal must be distinguished from that of man.

In his *Discourse on Method* René Descartes laid the foundation for a theory of language that recognizes and discerns linguistic acts as the representation of a subject. In order to do so he introduces a syllogism that pairs off humans, animals, and automata, and in effect uses the latter as a cultural technique capable of distinguishing humans from animals. The machine, in other words, is an anthropotechnique.

The first comparison reveals that from a philosophical point of view there is no difference between animal and animal machine. If there were a machine with the organs and the shape of an ape, "we would have no way of recognizing that they were not in every way of the same nature as these animals."[5] There are, however, two ways of distinguishing real humans (*vrais hommes*) from androids or replicants. The first is specialized language use. Only humans can put together words "as we do in order to declare our thoughts to others" (46). The second is reason. Automata do not act from insight or knowledge but according to the arrangement of their organs: "[T]hese organs need some particular arrangement for each particular action; hence it is morally impossible for there to be enough different arrangements in one machine to make it act in all the circumstances of life in the same way that our reason makes us act" (46). Reason is a "universal instrument" (46) that is always at our beck and call. To update Descartes' analysis, what animals lack is a Universal Turing Machine equipped with a freely programmable working memory. It is the limited number of possible ways of acting that reveals that machines do not act *par connaissance* (from knowledge) but only based on the *disposition de leurs organes* (the arrangement of their organs).

Descartes is staging a tournament. In the preliminaries, animals are up against animal machines. The result is a draw. Next, humans compete against human machines and manage to eke out a narrow victory. Machines are eliminated, and animals and humans move on to the finals. Now, as the two face off against each other, Descartes homes in on the crucial difference between human and animals linked to the usage (or lack thereof) of language and reason. Only a human, no matter how dense and dumb, can string together words in ways that render his thoughts comprehensible. Yet the inability of animals to perform this feat has nothing to do with lacking the necessary organs. Consider, Descartes reminds us, magpies and parrots; they "can utter words just as we do, and nevertheless cannot speak as we do, namely, by showing that they think what they say" (46). When it comes to the ability to combine words in order to represent thoughts, speaking birds are not even on the level of "madmen" (46). Had its soul not been created in a very different way, a trained parrot would surely be able to speak as well as a very stupid child. The sounds produced by animals do not constitute an animal language. Animal sounds— barking, grunting, cackling, croaking—divide language along a line that crosses and cancels the anthropological difference between humans and animals. Just like certain human sounds, animal sounds are "natural movements that indicate the passions and can be imitated by machines as well as by animals" (47). In other words, while animals are not in possession of human language, they are capable of producing certain parts of it: the interjections and exclamations that are the physical expression of affects.

The difference between the linguistic capabilities of humans and magpies forms the basis for a double representation of human speech. First, the infinite number of combinations of words into sentences represents the sum total of all the situations and objects of discourse, thus testifying to the universality of reason as the guiding instrument of the human machine.[6] Second, eloquence represents the thoughts of the speaking animal called man.[7] Together both represent the commanding position of the *instrument universel* known as reason.

PLINY'S PARROT

At this point it is necessary to insert a few historical asides concerning the philosophy and politics of talking birds. Both Greeks and Romans were aware of the fact that parrots could speak, but they did not view this particular ability as a specific difference between humans and animals. Pliny the Elder writes in chapter 57 of book 10 of his *Natural History*:

> But above all, there are some birds that can imitate the human voice; the parrot, for instance, which can even converse. India sends us this bird, which it calls by the name *sittaces*; the body is green all over and only marked with a ring of red around its neck. It will salute emperors and pronounce the words it has heard spoken; it is rendered especially lascivious under the influence of wine [*in vino praecipue lasciva*].[8]

Pliny emphasizes three points. First, parrots can speak, period. This applies in equal measure to humans and gods. Second, by saluting the emperor, parrots acknowledge that imperial authority also extends to the natural realm. Third, these qualities culminate in a proclivity to wine and lascivity. Building on Aristotle's observation that inebriated parrots become "more saucy than ever,"[9] this particular passage in Pliny serves as the basis for the iconographic association of parrots with matters of love.[10] Other speaking birds are also linked to desire rather than to anthropological difference:

> The magpie is much less famous for its talking qualities than the parrot, because it does not come from a distance, and yet it can speak with much more distinctness. These birds love to hear words spoken which they can utter; and not only do they learn them, but are pleased at the task; and as they con them over to themselves with the greatest care and attention, make no

secret of the interest they feel. It is a well-known fact that a magpie has died before now, when it has found itself mastered by a difficult word it could not pronounce. This memory, however, will fail them, if they do not from time to time hear the same word repeated; and while they are trying to recollect it, they will show the most extravagant joy, if they happen to hear it. Their appearance, although there is nothing remarkable in it, is by no means plain; but they have quite sufficient beauty in their singular ability to imitate the human speech.[11]

Forgetting and speaking belong together, as do speaking and the communicative function of clothes. Speaking, in other words, is a form of adornment that compensates for inconspicuous plumage: Just like bright feathers, it serves to attract attention. If you have little to offer the eye, wrap yourself in sound. From a functional point of view, animal or rather bird speech is primarily geared toward seduction. It is an external stimulus upheld by the animal's forgetfulness. (Phrased the other way round, we could say that the meaningfulness of words solely rests on our remembering them.) But no matter what: For Pliny, speaking, pleasure, and seduction still belong together. The ability to speak while constantly forgetting language is the basis of a birdlike pleasure whose implicit reason is the fact that the function of this particular ability is not to communicate but to stimulate the opposite sex. In and for birds, the pleasure of words and the pleasure of love coincide.

DANTE'S MAGPIE

The fact that one can distinguish between humans and birds was known to European speech experts even before the days of Descartes. One such expert was Dante:

And if it be claimed that, to this day, magpies [pice] and other birds do indeed speak, I say that this is not so: for their act is not speaking [actus locutio non est] but rather an imitation of the sound of the human voice—or it may be that they try to imitate us in so far as we make a noise, but not in so far as we speak. So that, if to someone who said "pica" aloud the bird were to return the word "pica," this would only be a reproduction or imitation of the sound made by the person who uttered the word first. And so it is clear that the power of speech was given only to human beings [datum fuisse loqui].[12]

Dante's competition between the linguistic capabilities of humans and magpies is far from endowing subjects with the potential to create some kind of spontaneous and universal speech. The accompanying table, which arranges *homines* and *picae* as *interpretanda* and *interpretaments*, reveals the chiastic relationship between Dante and Descartes. The distribution of affirmation and negation reveals that in the course of the three centuries that separate Descartes from Dante, the *interpretaments* shifted from the side of the magpies to that of the humans, while the *interpretanda* moved in the opposite direction. Given that the human ability to speak is a gift from God and therefore not in need of any interpretation, Dante defines it by analyzing the speech of the magpie: It is nothing but *repraesentatio soni*, a representation of sounds. By contrast, Descartes provides a positive definition of human speech, but he has nothing to say of the magpie other than that it "cannot speak like us."

	homines	picae
Dante (1303–04)	*actus locutio* (act of speaking)	*repraesentatio soni* (representation of sound)
Descartes (1637)	*en témoignant qu'ils pensent ce qu'ils disent* (proving that they are thinking what they are saying)	*ne peuvent parler ansi que nous*(cannot speak like us)

BALDUS'S MAGPIE

And yet there were in Dante's times human beings who inhabited language much as did the magpies. Medieval legal experts referred to them as *nuncii*. Their way of speaking has been forgotten, or rather: It has been replaced by the way gramophones inscribe His Master's Voice into the heart of Nipper and other loyal canines.

The *nuncius* emerges as a clearly defined discursive function in the course of the twelfth century. By the beginning of the thirteenth century, Azzo's *Summae Institutionum* identifies what Descartes was to banish from the domain of human animals: a human speech whose force is based on the fact that it resembles that of the *pica*. "A *nuncius* is he who takes the place of a letter: and he is just like a magpie . . . and he is the voice of the principal sending him, and he recites the words of the principal."[13] As Descartes would have it, the nuncius is an automaton: He is completely determined by the "arrangement of his organs," that is, by the finite repertoire of speech acts he is capable of and which happen to be completely removed from his own intentions. As a speaking being, the nuncius is a machine made up of and programmed by textual artifacts.

The fact that from a Cartesian perspective the nuncius fully satisfies the criteria of a language automaton has to do with a fundamental shift in the meaning of representation that occurred around the beginning of the early modern age. The medieval conceptualization of *representare* did not see it as the substitution for something that is absent by means of acoustic or pictorial signs. Rather, it meant that something absent was made present by way of physical embodiment. This meaning underlies the statement of Pope Gregory the Great: "Where we cannot be present, our authority is represented by those we command."[14] A nuncius speaks like one obsessed; he "speaks in his lord's person, never of himself."[15] The magpie's *repraesentatio soni*, described by Dante as an "imitation of the sound made by the person who uttered the word first," is an embodiment. A comment by the Italian jurist Baldus de Ubaldis (1327–1400) on the birdlike nature of the messenger proves that in the days of Dante, magpies were capable of performing this particular meaning of *representatio*: "For just as a magpie speaks through himself, and not from himself, and just as an organ does not have a sound by itself, so a nuncius says nothing from his own mind or by his own activity, but the principal speaks in him and through him."[16]

As Konrad Braun noted in his work on legations, Baldus's teacher, Bartholus de Saxoferrato (1313–1357; a contemporary of Dante), had made a similar point: "And other nuncii, whom we make use of when negotiating with absent parties, are as it were nothing but organs or (as Bartholus writes) magpies, by means of which those who are absent communicate."[17] The voice of the absent sovereign is to be heard in the voice of the nuncius. The communicative truth index rests less on the perceived intentionality of communication than on the presence of the noble body in a *pars pro toto*. The messenger is a kind of avian mask of the present/absent sovereign, a metonymically advanced secondary body. The principal resides behind the twittering of his messenger just as he resides behind the lion or donkey on his coat of arms.

HERDER'S PARROT: *LOGOS* AND *PHONÉ*

Yet it is not enough to draw a distinction between human and animal *logos*. The same operation has to be performed on the material into which the *logos* is inscribed: the *phoné*. The anthropological difference is not only of concern to a philosophy striving to safeguard the exclusive humanity of the *logos*; it is of equal concern to a pedagogy which no less zealously monitors the humanity of human *phoné*. The concern is warranted, for time and again human *phoné* insidiously thwarts the ontologically grounded exclusion of the animal from the human. What was said in the two preceding chapters about the information-theoretical distinction between symbol and signal, or the distinction between sign and channel, the transcendent meaning and the material dimension of a shared meal, also applies to the anthropological difference. The *ontological* difference between human and animal speech presupposes an *ontic* cultural technique of this distinction. In other words, the categorical exclusion of animal sounds from language needs to be enforced by practical, that is, disciplinary, measures.

Johann Gottfried Herder, who as General Superintendent and Head Consistorial Counselor of the Duchy of Weimar was heavily invested in secondary-school reform, led a two-front war against the animal sounds that Descartes had marked as the other of human language but that nonetheless could not be excluded from human speech. In Herder's paper from 1800 titled "Vitae non scholae discendum" ("Learn for life, not for school"), the parrot makes a topical appearance. Just as philosophy cannot do without the parrot, pedagogy is prone to summon it in order to help answer the question: What does it mean to learn? In the old education system, described in great detail by Heinrich Bosse, to learn was to imitate and recombine words that are not your own.[18] These words, however, are endowed with an almost physical force: "Words are sounds; they occasionally impress themselves on the youthful mind without any thoughts attached; but if learned in such a thoughtless manner one has acquired these words like a parrot that, as is well known . . . learns word-sounds which it reproduces at opportune and inopportune times."[19] The space of the old learning, the space of the Scholar's Republic, is a space of parrot discourse because it is one in which the words of others remain, as it were, alien words. And they remain alien because this space of learning is a space of words that belong to nobody, least of all to those who write them down. The space of learning is a space of discursive trafficking: Words move through and between scholars who are, without exception, secretaries and parrots of the Scholar's Republic.

But despite the necessity to exclude the mimicking parrot in order to purify the inward realm of Herderian man,[20] the sounds produced by the students in Weimar are still mired in the animal kingdom. Clearly, they do not live up to Herder's standard of human purity. Even when demonstrating the—according to Herder—uniquely human ability to express oneself using one's own words, the students are still not able to perform the Cartesian *logos*. A further cleansing operation, this time on the level of *phoné*, is required, because the *logos* appears to be in need of a specially constituted matter or mother in order to express itself. In 1796 Herder delivers the talk "Von der Ausbildung der Rede und Sprache in Kindern und Jünglingen" ("On the Education of Students in Language and Speech"), which for a century was hailed in German-speaking countries as the founding document of German as a school subject:

> When we come into the world we are of course able to scream and cry, but not to talk or speak; we emit only animal sounds. These animal sounds remain with some people and races throughout their lives. One has only to stand at a distance from which the sound of the voice and accent can be heard but not the meaning of the words: in some people one will hear the turkey, the goose, the duck; in others it will be the peacock, the bittern; and in pretentious dandies it will the canary; it will be anything but the human voice. Our province of Thuringia has many good things, but pleasant speech is not one of them, as one is prone to realize when one hears sounds, sounds mixed together, without understanding the meaning of what is said. Youths who have acquired this unpleasant dialect of merely animal sounds, be they from city or country, have to make every effort in school to acquire a human, natural speech possessed of character and rid themselves of their rural or shrieking back-alley dialects. They must leave off the barking and yelping, the clucking and cawing, the swallowing and merging of words and syllables and speak a human rather than an animal language. Happy is the child, the youth, who from his first years onwards hears comprehendible, human, sweet sounds, that imperceptibly mold his tongue and the sound of his speech. Happy is the child whose caretaker, mother or siblings, relatives, friends and finally first teachers speak to him in their bearing and speech with reason, decorum, and grace.[21]

Herder is operating with a double distinction between human and animal. First, there is clear distinction between parrots and students. Second, Herder effectively re-enters the human-animal distinction within the latter. Dialects are not compatible with the human in the full sense of the word. Because they are composed of animal sounds, dialects require a pedagogical hominization of the individual.[22]

HERDER'S SHEEP

It is no coincidence that Herder places animals at the starting point of humanity's *collective* entry into language, yet seeks to banish them from the beginning of *individual* language acquisition. The founding document of the anthropology of language, Herder's "Treatise on the Origin of Language," has human language, that stipulated Other of high-school animal sounds, emerge from the naming of an animal—a name, moreover, prompted by the animal itself.

"Already as an animal, the human being has language."[23] Herder's opening sentence explicitly revokes Descartes. "All violent sensations of his body, and the most violent of the violent, the painful ones, and all the strong passions of his soul immediately express themselves in cries, in sounds, in wild, unarticulated noises. A suffering animal, as much as the hero Philoctetes, when overcome with pain, will whine!, will groan!" (65). Descartes had warned against confusing such natural expressions of life with language. But this is precisely what Herder does: "These groans, these sounds, are language. Hence there is a language of sensation which is an immediate law of nature" (66).

The scene that Herder conjures up to depict the beginning of human language encompasses nothing less than the entire historical metamorphosis of anthropological difference from hunter, predator, berserker, and wolfman all the way to the domestication of herd animals. The creature that teaches humans to speak is not a bird but a very different young animal that, not coincidentally, belongs to the very first domesticated species: the lamb.

> Let that lamb pass before his eye as an image—[something that happens] to him as to no other animal. Not as to the hungry, scenting wolf!, not as to the blood-licking lion—they already scent and savor in their minds!, sensuality has overcome them!, instinct impels them to attack it! Not like the aroused sheep-man,[24] that feels the [she-]lamb only as the object of its pleasure, and which is hence again overcome by sensuality and impelled by instinct to attack it. . . . Not so to the human being! As soon as he develops a need to become acquainted with the sheep, no instinct disturbs him. (88; brackets in published translation; translation emended)

As Herder would have it, the end of foraging is the precondition for the acquisition of language. Herding is the antecedent of domestication. Paleolithic foragers are known to have organized battues or wildfires to drive herds of horses and elephants into swamps or pits.[25] One hunting technique in particular facilitated the transition from hunting to herding and thus to sedentarism: the pen or corral. Initially it served to capture wild goats. Arabian rock drawings in Mafraq, Jordan dating from the first millennium B.C.E. show rectangular and round enclosures into which goats are driven (Figure 3-1).[26] Rock drawings from the late Ice Age in Font de Gaume (France) and La Pileta, Malaga, depict similar pens, some of them with sketches of goats' heads or tracks instead of the captured (or kept?) herds. They precede the first domesticated animals in the Middle East by five to six thousand years and attest the old tradition of corralling that gradually transitioned into breeding.

It is the pen that keeps humans from attacking sheep in the manner of wolves and lions. Thus it is the pen that first inscribes into Herder's "sheep-man" the

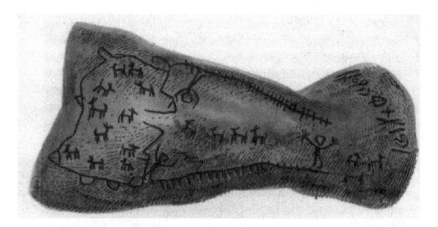

FIGURE 3-1. Rock drawing of a goatpen, from Mafraq, Jordan, first millennium B.C.E. Reprinted from Brentjes, *Die Erfindung des Haustieres*, 28.

difference between sheep and man (significantly, it remains unclear whether Herder's "sheep-man" refers to a ram or a shepherd). The sheep-man impelled by instinct to assail sheep is a monster, a humanimal, a being in which the difference between animal and human is suspended. It is the pen that first allows for a theoretically reflected relationship between man and sheep, an intellectual rather than instinctive relationship. Theory begins with the corral.

> [N]o sense tears him too close to the sheep or away from it; it stands there exactly as it expresses itself to his senses. White, soft, woolly—his soul, operating with awareness, seeks a characteristic mark—*the sheep bleats!*—his soul has found a characteristic mark. The inner sense takes effect. This bleating . . . remains for the soul.
>
> The sheep comes again. White, soft, woolly—the soul sees, feels, takes awareness, seeks a characteristic mark—it bleats, and now the soul recognizes it again! "Aha! You are the bleating one!" the soul feels inwardly. The soul has recognized it in a human way, for it recognizes and names it distinctly. . . . *With a characteristic mark therefore?* . . . "The *sound* of bleating, perceived by a human soul as the distinguishing sign of the sheep, became, thanks to this determination to which it is destined, the *name* of the sheep." . . . [W]hat is the *whole of human language* but a *collection of such words?* . . . [H]is soul has, so to speak, bleated internally. . . . Language is invented! ("Treatise on the Origin of Language," 88–89; emphases in original)

The first given name does not, as claimed by Kittler,[27] contain the difference between desire and the capacity to speak, but the culture-technical difference between prey and pets. In other words, the first name emerges from the difference between animals that are hunted and those that are kept; it is based on the difference introduced by corrals and pens. Herder tacitly posits a pastoral *nomos* as the precondition for the origin of language. The difference between natural, ovine bleating and its "onomatopoietic" repetition can only take effect once an obstacle is in place; it is opened up by the transition from prey to pet. The corral that excludes "desire" or "instinct" from sheep-shepherd relations serves to include animals in the phylogenetic origin of language. At the same time, it becomes necessary to devise cultural techniques aimed at excluding animals from the ontogenic origin of language, that is, from individual language acquisition.

Within the confines of the discourse network 1800, however, the function of the corral to split up the sheep-man into sheep and man is taken over by the Mother. It is not the domestication of the prey that brings about the hominization of humans but their pedagogical disciplining. At least this is what emerges from Johann Heinrich Pestalozzi's "Über den Sinn des Gehörs in Hinsicht auf Menschenbildung durch Ton und Sprache" (On the meaning of hearing with a view towards the formation of man by sound and language; 1804/1808). A prominent advocate of the new matricentric pedagogy, Pestalozzi exclaims:

> O Mother!—to whom I speak—once your child recognizes your voice as your own, its circle of understanding will expand further and further; it will gradually come to recognize the connection between birds and birdsong, dogs and barking . . . [28]

The Mother's voice grants the metonymic connection between the twittering sound and its magpie source. This transcendental voice rules the names of the animals much as the corral had once done; it rules over the split of desire and language that had not yet affected Pliny's parrot. And once this pedagogical disciplining becomes part and parcel of our primary socialization, psychoanalysts are able to discern the image of Mama behind cats and dogs.

FLAUBERT'S PARROT

In 1877 the parrot finally returns to the place from which it had been excluded by the many theories of language origin from Dante to Herder. This return occurs in Gustave Flaubert's "A Simple Heart."

"For half a century, the housewives of Pont-l'Evêque envied Madame Aubain her servant, Félicité."[29] The tale of the poor, uneducated housekeeper Félicité is a passion narrative. A paradigm of cleanliness, frugality, piety, and reticence, Félicité spends a lifetime of exemplary modesty in the service of others. After wasting her devotion on a brutish fiancé who deserts her, she attaches herself to her mistress's children, her nephew, and an old man with a cancerous arm. One by one they abandon her: They die, leave, or simply forget her. It is an existence that understands neither itself nor what is happening to it; a life in which an increasingly smaller number of love objects is forced to make up for an increasingly larger number of lost objects. "The final object in Félicité's ever-diminishing chain of attachments is Loulou, the parrot."[30]

Loulou is the gift from a baron, a former consul in America. Félicité sets out to teach the bird. Soon it is able to repeat: "Your servant, Monsieur! Hail Mary full of grace!" (45). Then, as a consequence of an infection, Félicité becomes deaf.

> The little secret of her ideas grew even smaller, and the peal of the bells, the lowing of the oxen, did not exist any more. All beings functioned with the silence of phantoms. One single noise reached her ears now, the voice of the parrot. As if to amuse her, he would mimic the clicking of the roasting spit, the shrill cry of a fish vendor, the saw of the carpenter who lived across the way; and, at the sound of the doorbell, imitated Madame Aubain: "Félicité! The door! The door!" (48–49)

The parrot is all Félicité can hear, and it imitates all she can no longer hear. The sounds it produces replace the inaudible original sounds. The parrot turns into an acoustic simulation medium. As the very last love object, Loulou comes to occupy the transcendental position once held by the very first love object, Pestalozzi's mother. While Romanticism was able to detect the poetry of nature—that is, the voice of the mother—at work in the sounds of the world, the world audible to Félicité is produced by her parrot. Parrots no longer imitate, they reveal the world.

And then, just like that, Loulou dies. But because there is no successor able to attract Félicité's devotion, she has him stuffed. If you have been robbed of everything, only images remain. Loulou is placed on a small shelf over a fireplace that juts out into the room. As an effigy Loulou first transcends the divide between life and death, then the boundary between thing and sign, and finally the distinction between divine pneuma and animal speech.

> At church, she would always gaze at the Holy Ghost, and noted that it had something of the parrot about it. Its resemblance seemed even more obvious to her on the brightly colored Epinal lithograph that represented the baptism of Our Lord. With its purple wings and emerald body, it really was the spitting image of Loulou. (53)

While parrots and Holy Spirits can speak, doves cannot. Félicité's simple logic corrects the mistake in the Christian ornitho-theological pictorial program. "The Father, to express himself, had not chosen a dove, since those animals have no voice, but rather one of the ancestors of Loulou" (54). Is there a model for Flaubert's replacement of dove by parrot? Indeed there is.

The Greek word for the breath or wind with which the Holy Ghost comes over man is *pneuma* (πνεῦμα). And since the Holy Spirit is nothing but a pigeon carrier channel, it, too, is called *pneuma*. The study of the Holy Spirit, then, is a form of pneumatics, and the encounter between parrots and pneumatics takes place in the natural science cabinets of the Enlightenment, as depicted in Joseph Wright of Derby's famous 1768 oil-on-canvas painting *An Experiment on a Bird in the Air Pump* (Figure 3-2). The actual experiment was probably performed by James Ferguson in 1762. Ferguson (1710–1776) was a precision-instrument maker who traveled through England to give public lectures on experimental philosophy. It is known that as of 1756 his equipment included an air pump. In 1763 Wright himself assisted him in his air pump experiments.[31]

But why did Wright choose of all birds possible a parrot—more precisely, a white cockatoo (*Cacatua sulphurea*)? After all, parrots were not common in

FIGURE 3-2. Joseph Wright of Derby, *An Experiment on a Bird in the Air-Pump*, 1768. Oil on canvas, 183 × 244 cm. © The National Gallery, London.

eighteenth-century Britain. When Robert Boyle and James Ferguson placed birds inside the air pump, they tended to recruit native species such as sparrows, larks, or ducks. By contrast, the white cockatoo was such a rare bird that its first description dates from 1760.[32] So how did it end up in the painting? The art historian Werner Busch found the solution in the uncanny likeness between Wright's work and depictions of the Trinity found in Dutch paintings.[33] Wright's parrot has taken over the position occupied by the dove-shaped Holy Spirit in his Dutch models. The *tertium comparationis* underlying this substitution is the speaking with nonhuman tongues. The parrot's capacity to speak, which enabled it to become part of the menagerie of Descartes' philosophical animals, and the fiery tongues of the Holy Spirit are the two nonhuman poles framing all the theories of human language. Wright combines pneumatics with pneumatology.[34] The Pentecostal miracle of speaking in tongues resonates in the parrot's prattle; and precisely this resonance marks Félicité's religious practice. None of this is coincidental, for Flaubert was acquainted with Wright's vacuum parrot. As his workbook no. 13 indicates, he saw the painting in London in July 1865 during his travel to Britain.[35] Flaubert's ornithology effectively merges the two manners of speaking that in the course of Western history and civilization had been pushed to the opposing extremes of bird speech and glossolalia.[36] Loulou unites the medieval function of the *nuncius*, the Cartesian definition of the bird as a language machine, and the revelatory function of the Holy Spirit, which constitutes the opposite of merely imitative speech.

Flaubert, who famously left nothing to fantasy, had borrowed a stuffed parrot from the Rouen museum when he began to write *A Simple Heart*.[37] At this point it is tempting to follow the lead of Julian Barnes and speculate on a possible identification between Flaubert the writer and his parrot—an identification, Barnes adds, that may have been fueled by Buffon's observation that parrots are prone to epilepsy. "Flaubert knew of this fraternal weakness: the notes he took on parrots when researching *Un coeur simple* include a list of their maladies—gout, epilepsy, aphtha and throat ulcers."[38]

But another source sheds more important light on Loulou: the extensive dossier compiled by Flaubert for the second, unfinished part of *Bouvard et Pécuchet*. As Flaubert wrote in the year of his death to Madame des Genettes, it "will consist almost entirely of quotations"[39]—that is, of verbatim repetitions of what the protagonists were copying. Among these copies is a newspaper clipping from *L'Opinion nationale* of June 20, 1863, about a man who had revered his parrot as a saint and who following its death fell under the delusion of being a parrot himself.[40]

The fact that Flaubert had marked this story for the second part of *Bouvard et Pécuchet* suggests that the parrot was to act as an emblem of the two clerks.

Its speech, made up exclusively of repetitions, is an acoustic pendant to their copying activities. In contrast to the labor of poets, the work of clerks consists of processing knowledge on the level of its literal materiality. Its hallmark is counting rather than recounting, registration as opposed to understanding. Starting in 1872, the year in which he began work on *Bouvard et Pécuchet*, Flaubert's workbooks are filled with long lists enumerating the books he read month after month. What Flaubert did in the last eight years of his life—and to a lesser extent while working on *Madame Bovary*, *Salammbô*, and the *Temptation of Saint Anthony*—is precisely what Bouvard and Pécuchet end up doing: copying. Evidently, Flaubert's authorship is on the verge of being swallowed up by the cultural technique of his protagonists. The work of the clerk, for whom knowledge exists solely under the aspect that it is composed of letters in need of copying and counting, is the media-archeological point at which the signifier returns to itself in order to denote its own origin or *arché*. "They return to copying. And to them, copying means to write *Bouvard et Pécuchet*."[41] Content and production of the novel coincide, which is why it had to be interminable. Flaubert was doomed to die over (and of) *Bouvard et Pécuchet*.

Flaubert's project is no grammatology. Operating on the level of actual speech, it is concerned with the brute fact that statements have been made rather than with the more abstract rules determining how and whether they make sense. Its intention is to infect speaking beings with the virus of the *déjà lu*. The *Dictionnaire*, Flaubert wrote, has to be designed in such a way that "qu'une fois qu'on l'aurait lu on n'osât plus parler de peur de dire naturellement quelque chose qui s'y trouve."[42] Because one can "naturally" not say anything that has not already been printed—in other words, because one cannot say anything without saying it as a parrot—one cannot say anything at all.

In his later years Flaubert's poetics were marked by the gesture of willingly forfeiting the gift of anthropological difference. As the writer turns into a parrot, humans are transformed into channels directly linking god and animal. Flaubert returns the difference between bird speech and human language, thereby exchanging his role as author with that of secretary or parrot. A parrot that reveals human speech as an assemblage of commonplaces, a variant of parrot discourse, which deserves to be silenced. A parrot as the truth and end of all speech.

4

MEDUSAS OF THE WESTERN PACIFIC

The Cultural Techniques of Seafaring

ANTHROPOLOGY OF SEAFARING

What is a ship? Physics provided an early answer courtesy of Archimedes: A ship is a body that floats in water because it displaces its weight in fluid. *Heureka.* Naval architecture, which speaks of ships only in the plural—galleys, carracks, outrigger canoes, three-deckers, herring luggers—offers a more differentiated response: The ship is a colonial and scientific technology, a medium of overseas trade or ceremonial exchanges, a war machine, a technology for harvesting fish and other ocean riches. It is, no doubt, a very protean first-order technique, yet it is more than that. As Foucault notes, it is "for our civilization . . . the greatest reserve of the imagination."[1] Ships do not only transport humans and goods, they also carry innumerable stories. Myths, holy scriptures, literary texts, and films all invoke ships to tell tales of departure, failure, and struggles with the elements, of encounters with other laws, the law of the other, and that which is outside the law. So, once again, what is a ship? Third response: "Un vaisseau est toujours un résumé parfait de l'espace comme il est."[2] The seamen among the philosophers and the writers among the seamen (think of Herman Melville and Joseph Conrad), the connoisseurs of political topography and designers of political imagery, have always depended on the representational qualities of the ship as a second-order technique.[3] Inevitably, the question of what a ship is requires that we deal with the ship as simile, metaphor, or emblem. The journey of life and the passage into the land of the dead, the church, the state, but also madness and the notion of "navigating" the internet, are firmly linked to the ship and its parts, be they sails, anchor, or rudder.

On the other hand the ship is also a summary of the world as it is not. "The ship," to return to Foucault (and this constitutes the fourth response), "is the heterotopia *par excellence*,"[4] because it combines, as it were, all the different forms of heterotopia enumerated by Foucault. As a ship of fools loaded with madness, or as an eighteenth- or nineteenth-century training vessel enclosing the adolescence of young naval cadets, the ship is a crisis heterotopia. By offering asylum to idlers, criminals, and future pirates, ships are heterotopias of deviation. Like the garden, the ship "is the smallest parcel of the world and then it is the totality of the world."[5] By traveling from harbor to harbor (or brothel to brothel), it is a heterotopia of illusion; and by confronting the chaotic and disorderly outside with a perfectly ordered world in which every item is assigned a precise place within a complex arrangement, and each action takes place at a precise moment within an equally complex procedure, it is a heterotopia of compensation. Heterotopias exceed topos, image, or summary: They are different yet nonetheless concrete spaces that mirror, invert, connect themselves to, or sever themselves from the world as it is.

To explore the ontology of ships, then, requires an anthropology of seafaring, given that the latter appears to be ground for the metaphorology of the ship. Humans may be creatures of dry land, but the comparisons they use to illustrate their existence are recruited from a domain of being to which they are not native: the sea and the cultural technique of seafaring.[6] In other words, in order to deal with the ship as an image of excess meaning, a summary and heterotopia of the world, we need to address seafaring as a human activity that challenges and confronts the expanse of maritime space. However, what humans do with ships matters less than what seafaring does with and to them:

Numberless wonders
terrible wonders walk the world but none the match for man—
that great wonder crossing the heaving gray sea
driven on by the blasts of winter.[7]

The first stasimon of the Theban choir in Sophocles' *Antigone* extols the *chorein* of those who leave the *polis* to travel the gray seas: "Poliou péran póntou . . . chorei." *Chorein* is the verb form of *chora*. Plato's *Timaeus* describes the latter as a third form of being next to the unchanging and immutable ideas (*eidos*) and the constant flux of the perceptible world:

And the third type is space [*chora*], which exists always and cannot be destroyed. It provides a fixed site [*topos*] for all things that come to be. It is itself apprehended by a kind of bastard reasoning that does not involve sense perception, and it is hardly even an object of conviction [*pistis*]. We look at it as in dream when we say that everything that exists must of necessity be somewhere, in some place [*topos*] and occupying some space [*chora*].[8]

As Heidegger emphasized, *chora* is not to be mistaken for space in the sense of *extensio*. "The Greeks have no word for 'space.' This is no accident, for they do not experience the spatial according to *extensio* but instead according to place (*topos*) as *chōra*, which means neither space nor place but what is taken up and occupied by what stands there. The place belongs to the thing itself."[9] *Chora* is that which makes space so that something, a thing, may appear. But it is also the outside of the *polis*. *Choreo*, the verb, means "to make place" or "give room," but it also translates as "passing through, penetrating, traversing successfully."[10] The ship makes space. By setting out and carrying man, "that terrible wonder," into open waters, it transforms the sea, hitherto devoid of any sense of place or history, into something inscribed by both. It marks the *chora* with spatial and temporal difference and thus creates the possibility of controlling maritime space, founding colonies, and visiting distant shores to taste and trade new wine or subjugate alien tribes. It is this act of creating space within emptiness that turns humans into the most uncanny of creatures—cultural beings. Successful seafaring is forgetfulness of the sea. By turning the sea into a zone of intercultural contact, a space in which all kinds of histories of war, economy, and culture may unfold, the *chorein* obscures what this historical space emerged from. Only when we suffer shipwreck do we repay what we owe the *chora* for transforming the sea into a space of human history. Ultimately, shipwreck is inevitable. After all, it is not just any sea the Sophoclean *anthropos* dares to navigate, but the winter sea churned by heavy gales; it is the very sea that Hesiod's *Works and Days* declares to be off limits:

> For fifty days after the solstice, when the summer has entered its last stage, the season of fatigue, then is the time for mortals to sail. . . . Then without anxiety, trusting the winds, drag your swift ship into the sea and put all the cargo aboard. But make haste to come again as quickly as you can, and do not wait for the new wine and the autumn rains, the onset of winter and the fearsome blasts of the South Wind, which stirs up the sea as it comes with heaven's plentiful rains of autumn, and makes the waves rough.[11]

If, as Sophocles implies, the creation of space begins with the dangerous voyage driven on by the winds of winter, with leaving the *topos* or *polis* behind by venturing out into the time- and spaceless gray expanse, then the ship precedes even writing as an original cultural technique of hominization. For Heidegger, the *choreo* of seafaring appears to belong to the very origin of ontological difference. It is something of which one cannot speak—only sing. Seafaring is what enables men to commit all the monstrous offenses against the goddess Gaia described by the Sophoclean chorus.

TROBRIAND NAUTICS

The archives of Western culture contain an abundance of knowledge related to the cultural technique of seafaring. Nonetheless, occidental culture cannot provide a satisfactory response to the question of what a ship is, because the oldest European literary and archaeological evidence presupposes the topical contrast between ship and ocean. Our rationality always already separates religion and shipbuilding, literature and navigation, into two different types of knowledge: culture on the one hand, technology on the other. Clearly, this doesn't settle matters. For the seafaring cultures of Europe, the knowledge of what makes a ship seaworthy and the knowledge of the terrors and dangers awaiting those who sail the seas are fundamentally separate issues. It would not even help to discover seafaring sagas that precede the story of the Argonauts, given that Greek mythology offers no more than hints of an original coupling of ship and sea. Instead we need to head towards non-European waters and study the nautical practices of non-European Argonauts. It was not by chance that Bronislaw Malinowski called his epoch-making book of 1922 about the Trobriands *Argonauts of the Western Pacific*, despite the fact that it is not about Greek heroes but the inhabitants of a Melanesian island world, the Massim, who travel great distances with their canoes.[12] Ever since Malinowski's book the Massim, that is, the inhabitants of the archipelagos east of Papua New Guinea, have enjoyed considerable fame not only among ethnologists but also among scholars of cultural studies. The reason, however, has less to do with the way they manage to sail the seas than with the ceremonial ring-exchange of necklaces and armbands they practice. Known as the Kula, it amounts to a socioeconomic system of exchange among fourteen island groups separated from one another by distances ranging from 50 to 170 kilometers. Nevertheless, the outrigger canoes used by the Massim are an indispensable precondition of the Kula, which is why Malinowski ascribed a central role to Melanesian seafaring within the social organization of the Trobriands. The canoe is a machine that produces not only the Kula exchange but also the maritime society of the Trobriands. It is a network comprising many highly heterogeneous actors, including creepers, myths, several kinds of magic, taboos, flying women, and the threat of shipwreck.

To sail across the ocean by means of an outrigger canoe means above all one thing: to be confronted with the horrors of the sea. Of these, the flying witches (*mulukwausi*) are the most terrifying. The Trobriands believe that there are many witches (*yoyova*) among the women of the tribes that live on the islands surrounding them, especially on Kitava, Dobu, and the Amphlett Islands. A woman who is a yoyova can send out a doppelganger (*mulukwausi*) that is able to fly across the ocean and either make herself invisible during her flight or assume the shape of a night bird or a firefly (238). The main occupation of the mulukwausi is to feed on dead bodies and kill sailors in distress. If a canoe is in danger the witches will hear the cries for help and fly to the scene of distress. "The sea and sailing upon it are intimately associated in the mind of a Boyowan [an inhabitant of Kiriwina, the main Trobriand island] with these women" (237), It is therefore imperative to understand the role played by these witches in order to grasp the relationship between the ship as a technical object and the sea as a space filled with demonic horrors.

During the first stage of the building process, as he is hollowing out a tree, the canoe builder performs an extensive magic rite designed to secure the highest speed possible for the canoe. The Kula is a competitive affair, hence it is not only important to get hold of the most precious necklaces but also to sail the fastest canoe. With the latter in mind the magician recites a long spell to ensure that this particular canoe will leave all others behind. The spell ends with the words *saydididi*, which imitates the sound the witches make during their flight. But the canoe is identified with the flying witches even prior to its actual construction. Before it is used to hollow out a trunk, the ax is subjected to a magic rite (*ligogu*) that prompts the future canoe to fly: "Bind your grass skirts together, O canoe . . . fly!" (132). This corresponds exactly to the way in which the witches are said to fly: by tying their skirts tightly around their bodies (242). In the course of the *wayugo* magic, which is performed during the second stage of the construction process and which is equally decisive for the vessel's seaworthiness, the magician once again addresses the canoe several times: "Thou, o my boat, bind thy skirts together and fly!"(138). *Wayugo* is the name of the creeper used to lash the rib framework and clamp together the canoe's other parts. Since the cohesion of all the parts of the canoe and thus the survival of the sailors at sea depends on the strength and proper application of the creeper, the magic pertaining to it is of the utmost importance for the proper operation of the vessel. Both the ligogu and the wayugo spells contain a part that predict a certain route: "[B]reak through your sea-passage of Kadimwatu, cleave through the promontory of Saramwa, pass through Lomu; die away, disappear, vanish with an eddy, vanish with the mist" (138–39). This part of the canoe magic alludes to the Kudayuri myth, which is of fundamental significance for the connection between canoes and flying witches. Kudayuri is a village on the island of Kitava that can be seen just over the horizon east of Kiriwina. Here once lived Mokatuboda together with his younger brother Toweyre'i and their three sisters. They all had come out of the earth at this spot and were the first to possess the ligogu and wayugo magic (the magic essentials for canoe building). When the men of Kitava decided to prepare an extensive Kula expedition, Mokatuboda also built a canoe, but in ways that contradicted the basic construction common among the Trobriands. Instead of taking the canoe to the beach to lash together its different parts, Mokatuboda left the canoe inside the village and accomplished the task without the participation of the rest of the community. Mokatubodu could afford to deviate from the established ritual because he knew the magic to build a canoe that flies—which ensured that he was always the first to reach the next stop of the Kula expedition although he was inevitably the last to depart. Toweyre'i, his younger brother, in the mistaken belief that he had acquired Mokatuboda's magic construction skills, killed his sibling together with his male matrilineal relatives, but in the following year it turned out that he was not in possession of the required talent. Ever since, the art of building flying canoes has been lost. The three sisters got very angry with Toweyre'i because he had killed their brother for nothing. However, they were yoyova who had learned the ligogu and wayugo magic by themselves; turning themselves into mulukwausi they left Kitava as flying witches and flew across the air as Mokatuboda's canoe had done before. One flew to the west, while the other two headed south in the direction of Dobu, thereby creating the strait to the south of Kiriwina flanked by the islands of Tewara and Uwama on the one side and Dobu and Ferguson on the other. This is the strait through which the Trobriands pass on their southward Kula expeditions.

The myth tying the origins of the mulukwausi directly to a legendary flying canoe thus contains a kind of "portolan" referring to geographical landmarks most important for Kula navigation.[13] The Kula magic transfers the ability of the mulukwausi to fly to the canoe; moreover, it identifies the canoe with a flying witch and thus addresses it as part of the very horror of the sea most feared by the Kiriwina sailors. Before the fratricide there was a flying canoe but there were no flying witches (and no horrors at sea); after the mythical crime there are flying witches and the sea routes of the Kula but no flying canoe. Instead, each canoe built on the island is invested with a velocity that originates in the magical forces of the witches. The myth of the fratricide appears to provide an explanation why it is necessary for the canoe magic to make a pact with the powers of evil and the horrors of the surrounding seas in order to ensure the necessary speed and safety. These powers and horrors emanate from women. Since the mythical loss of the exclusively male power to make canoes fly, sailors have to make do with a power derived from females to make their canoes fly through the water. This potency, no longer the exclusive possession of a single magician, can be activated by every canoe builder on the island, but it is now of female origin and comes from outside; it belongs to the power of the sea and its daughters. Without that potency, which looms on the islands behind the horizon and spreads fear among men, no canoe will ever find safe and speedy passage across the ocean. This is something Homer never dreamed of: that the sirens not only threaten the ship of Odysseus but render it seaworthy in the first place. This dialectics is the essence of the *apotropaion*. The ship has to become a siren; it has to transform into a flying witch.

A SEA OF GAZES

Malinowksi indicated that a "definite connection" (244) existed between the mulukwausi and the many dangers one may encounter at sea—be it the "gaping depth" (*ikapwagega wiwitu*), also prominent in ancient Greek horrors of the sea,[14] or such sea creatures as sharks, smaller sea animals, crabs, mussels, and other things. He did not explore this connection in depth, despite the fact that the stories and magic spells he collected make it more than clear. It is the gaze. The waters around the Trobriands are full of gazes; indeed, the sea itself gazes. Consider the magic of mist, *kayga'u:* It is the "indispensable magical equipment" (245) a sailor uses to protect himself against the danger of shipwreck as well as against "the omnipotence . . . of woman" (245)—which is practically the same thing. The canoe of the Trobriands is an object of incessant observation; it is being looked at from all sides with such intensity that the protective magic employed must consist in either rendering it invisible or blinding the gaze of sea. The principal effect of the kayga'u magic is to create a kind of mist that blinds the evil things capable of causing the death of a shipwrecked person. The mulukwausi that follow the canoe, the sharks and jumping stones that wait in ambush, the crabs, the poisonous or spiky fish, and the "gaping depth" itself all are warded off and blinded by the mist.[15] This implies that the evil impact of the flying witches, the gaping depth of the sea, and all the other terrifying things out there emanate from a gaze. What terrifies the sailors during the Kula voyages is the prospect of an evil eye looking at them from every possible vantage point. "I befog," intonates the magician practicing the kayga'u magic, "I befog the eyes of the witches! I befog the eyes of the little crabs! I befog the eyes of the hermit crab! I befog the eyes of the insects on the beach!" (250). The spell leaves no doubt that that the gaze is connected to the female gender: "I befog the eyes of the women of Wawela; I befog the eyes of the women of Kaulasi; I befog the eyes of the women of Kumilabwaga, I befog the eyes of the women of Vakuta!" (ibid.). The sea of the Trobriands is not only a sea replete with gazes, it is also one full of evil women.

What we have here is an instance of the evil eye in which the latter does not adhere to a single "strong women" pursuing her dream of omnipotence to be the only mother,[16] but one in which the evil eye resides in anything that can become dangerous for the sailor at sea: insects, worms, spiky fish, sharks, fire-flies, jumping stones, the very depths of the ocean. They all are different forms of one single object that, following Lacan, we can call *objet petit a*. Derived from the Freudian concept of "'a little one' that can become separated from one's body,"[17] it is born from an original separation, a self-mutilation induced by the proximity of the Real.[18] As Žižek puts it, the "consistency of our 'experience of reality' depends on the exclusion of what he [Lacan] calls the *objet petit a* from it."[19] In other words, to have normal access to reality, something must be excluded or originally suppressed from the latter. In the field of visibility the *objet petit a* is the gaze. Lacan inherited from Sartre the differentiation between eye and gaze. The gaze is always outside—where the classical order of vision localizes the objects. "I only see from one point, but in my existence I am looked at from all sides."[20] Because the gaze is able to symbolize the fundamental lack linked to the fear of castration, it is closely tied to desire and, reciprocally, to fear. However, what is not thematized in Lacan's discourse is that the prominent examples he uses in his *Seminar, Book 10* and *Seminar, Book 11* to illustrate the operations of the gaze originate in the maritime domain. In the *Seminar, Book 11: The Four Fundamental Concepts of Psychoanalysis*, Lacan recounts that as a young intellectual in the 1920s he would go fishing with a couple of fishermen in a small boat off the coast of Brittany. They ventured into the very same waters as the Breton fishermen in Jean Epstein's films. Neglecting the difference between the necessity of making a living and the intellectual pleasures of a leisurely pursuit, the young Lacan wanted to share with his fishermen the dangers Epstein's fishermen were forced to expose themselves to. While they were waiting until it was time to haul aboard the net, a fisherman called Petit-Jean pointed at a sardine can floating in the water and glittering in the sun. "And Petit-Jean said to me—You see that can? Do you see it? Well, it doesn't see you!"[21] As Lacan informed his audience, he did not find the story funny at all, because the can did in fact look at him, and thus signified the meaning of the words that Petit-Jean told him, namely, that it did not look at him in the sense that it did not concern him. "[I]t was looking at me at the level of the point of light, the point at which everything that looks at me is situated."[22] The gaze realized by the sunlight reflecting off the can destroys the mimetic image which the young Lacan had believed would enable him to become part of the ship's crew.

Lacan's second example for the gaze as an *objet a* is taken from the final sequence of Fellini's *La dolce vita*, when Marcello (Marcello Mastroianni) and his fellow revelers end up on a beach where he sees the motionless eye of a thing from the sea which the fishermen have just produced.[23] One of the girls calls it a "sea-monster"; for one of the gay dancers it is "the head of the Medusa." "Look,"

Marcello says, "how he is staring at us." That's it, Lacan says, that is that which concerns us all, that gazes at us all and thereby shows how the fear is coming into the visible at the place of desire.

Stories of the gaze are stories of the sea. However, what happened to Lacan in that fishing boat off the Breton coast is nothing compared to the sea of the Trobriands. The latter is a sea full of sardine cans. Of course it doesn't take a sardine can for a gaze to emerge from the sea. A firefly will do, a glistening light on the water at noon when the wind dies down, a rustling of the pandanus leaves, or a gargle in the depth. They all contain flying witches and gazes that force their potential victims to seek shelter by means of a protective screen.

Thus the voyages of the Trobriand Argonauts unfold and illustrate the Sophoclean notion of seafaring as a primary cultural technique. Insofar as shipwrecks attract the gaze of the flying witches that realizes the split within the subject, and insofar as seafaring is an unceasing deferral of shipwreck, seafaring is an anthropotechnique. Humans do not turn into sailors by setting out to sea; rather, the sailor, that most uncanny of all seafaring beings, is "d'homesticated" by the ship.

EVIL AND APOTROPAIC EYES

In 1954 Edmund Leach published a short paper on the war shields of the Trobriands, in which he interpreted their decoration as the folded representation of a flying witch or mulukwausi (Figures 4-1 and 4-2). Since the purpose of the decoration is to petrify the attacking enemies, Leach called the image of the mulukwausi a "Trobriand Medusa," thus establishing a connection between the real Trobriand experience and one of the most famous myths of the evil eye.[24] It, too, involves the sea. Medusa, one of the three Gorgon sisters, was once a beautiful women. One night Athena came across Medusa and Poseidon making love in one of her temples. In her anger she transformed Medusa into a terrible monster with glowing eyes, huge teeth, a lolling tongue, and snakelike curls. Her sight turned human beings into stone. After Perseus beheaded Medusa, Athena fixed her severed head on her aegis or, according to other sources, her shield.[25] The fact that the gorgoneion—the representation of the Medusa's head—was in archaic times a shield decoration corresponds to this version of the myth.[26]

By invoking the name of Medusa, Leach interpreted the Trobriand mulukwausi as a version of an anthropological universal. As a result, the apotropaic entanglement of ship and ocean achieves universal validity, and certain measures of protection documented by archaeological findings in the Mediterranean

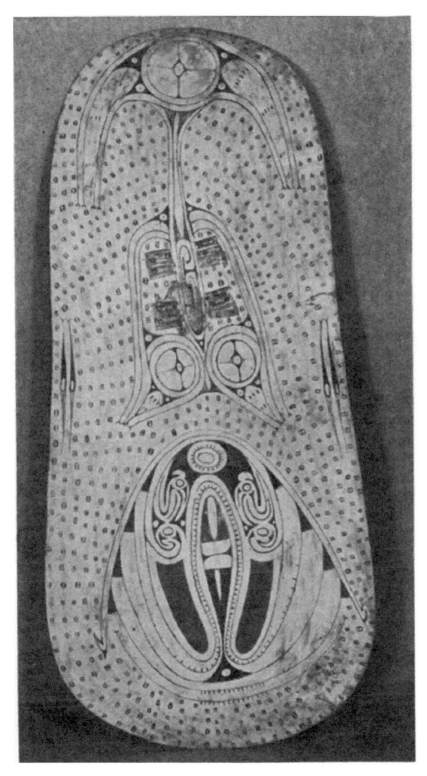

FIGURE 4-1. Shield from the Trobriand Islands. Reprinted from Leach, "A Trobriand Medusa?" 104.

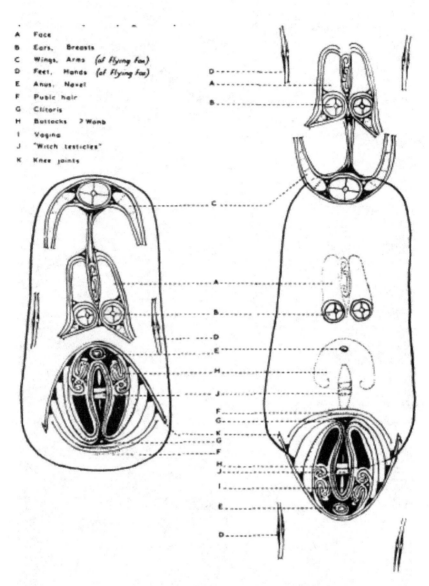

FIGURE 4-2. Designs on the Trobriand shield interpreted by Edmund Leach. Reprinted from Leach, "A Trobriand Medusa?" 104.

have to be interpreted as remains of a connection similar to the one Malinowski encountered among the Trobriands. In much the same way that a *masawa* canoe is armed with the eye of a flying witch in order to ensure speed and safety,[27] eyes were attached to the prows of Greek vessels.[28] As suggested by archeological findings in Zea (part of the port of Piraeus) and Tektaş Burnu,

Turkey, the eyes were not painted on, but made of marble. It is assumed that they served as a means of protection from the evil eye of supernatural creatures of the depths who, like Poseidon, envied the mortal seamen their return home to their happy life on dry land.[29] The marble eyes found in Zea are in all likelihood apotropaic devices from around 500 to 300 B.C.E.[30] The ships of the mysterious sea peoples of the Aegeis, who were defeated by Ramses III in a sea battle during the twelfth century B.C.E., already displayed eyes at their prow designed to ward off attacks from creatures of the depths, possibly sea dragons.[31] The discovery of a leaden anchor cross beam probably from Hellenistic times, decorated with the head of the Medusa, provides evidence that the ancient Greeks used the gorgoneion as an apotropaion against the evil that lurks at the bottom of the sea.[32] Because anchors are lowered to the depths where they come into immediate contact with the creatures residing there, they represent a threat to the ship when raised. Magic spells are necessary to protect anchor and ship from the evil eye of the depths. In his study of the evil eye Thomas Hauschild has amassed a lot of evidence supporting the view that the connection between Medusa and the sea is at the center of the Mediterranean complex of the evil eye. The Islamic Quarina, which like the mulukwausi is the evil "double" of a woman who died while giving birth, has a fish tail and comes from the sea.[33] Her gaze renders other women infertile and kills their progeny. The connection between gaze and sea is also apparent in the fish-tailed Neapolitan siren amulets designed to ward off the evil eye. As a rule, images of fish-tailed women in antiquity refer to Scylla, a sea monster with multiple heads snatching up unlucky sailors from the decks of passing ships. Other depictions show sirens entwining sailors with their tentacle-like tails.[34] During the Middle Ages sirens merged with fish-tailed mermaids.[35] But the gaze of the ambivalent woman, part saint and part evil double, may in turn transfix the sea itself. The "Eye of Santa Lucia" that adorns many Italian fishing boats promises a rich haul.[36]

A THEORY OF THE SHIP

Within the material culture of the Trobriands, shield and canoe are structurally opposed to each other. On the one hand both the painted war shield and the canoe—especially the creeper used in construction—are armed with the transfixing gaze of the mulukwausi.[37] On the other hand the shield is a male object, while the canoe is clearly female. The canoe exhibits this gender difference by oscillating between terror and beauty, depending on whether it is gazed at from the sea (by flying witches) or from land (by young women on the beach). The sailors on board a Kula canoe are convinced that they provide an irresistibly attractive sight for the girls and young women on land.[38] This particular feature completes our picture of the theory of the ship, as based on the seafaring practices of Western Pacific Argonauts—a theory that also reveals the systemic place occupied in modern European culture by the seascape. The fascination that grips the female spectators as they gaze out on the beautiful canoes is no less part of the canoe magic than the transfer of the powers of the flying witches to the canoe. In other words, the canoe realizes the narcissistic function of the tableau, insofar as the young men on board imagine themselves as irresistibly attractive images in the eyes of the beautiful women who watch them from the shore. They are making a *bella figura*. In the European culture of perception, this function was enhanced during the sixteenth and seventeenth centuries to such a degree that it came to dominate much of the use and appearance of ships. The magic of the ship to make its owner and crew appear as erotic superheroes turned into the capability to represent the glamour, the power, and the sovereignty of a prince or nation, while the entanglement with the evil eye sweeping around in the ocean was nearly completely excluded from the image.[39] When viewed from the land, the canoe is a seascape; when viewed from the sea, it is a Medusa's shield. The ship, then, operates between land and sea as a double being of image and gaze. It fascinates the gaze of the sea (that is, the gaze of the flying witches) in the apotropaic sense; and it exerts an erotic fascination on the eyes of the women on land. There is a clear dialectics between the desired effect of attraction on the land-based women and the feared attraction on the sea-women. Women are split into the duality of sirens and lovers, and this duality is articulated by the difference between land and sea. In comparison to this dialectic, attempts to show that images of death and corpses are at the origin of the anthropology of the image appear to be pacifying or belittling matters.[40] If the original entanglement of gaze and image is at stake in concrete and mythical situations of the evil eye, then the image is originally apotropaic. Images are not originally linked to corpses, but to objects armed with the very gaze they are designed to ward off. Aesthetically, the image appears thus as a threshold that mediates between shape and shapelessness, wholeness and disunity, inclusion and exclusion of the *objet petit a*, the imaginary and the real. The ship, therefore, has always been image and gaze: a gazing image that wards off the gaze of the ocean, an image that fascinates the eyes on land. The Trobriand Islands present us with a theory of the ship that systematically combines a theory of the sea and its horrors, a theory of the gaze and the image, the problems of gender difference, the difference that is the female gender in itself, and a theory of nautics as anthropotechnics. The sea is a domain of reality in which the exclusion of the *objet petit a* did not take place.

Instead it is included within reality, which subsequently entails that seafaring may cause a disintegration of reality.[41] The representation of the sea therefore makes aesthetic as well as ontological sense. In order to represent the sea, it is necessary to exclude the gaze from it. In order to prevent the disintegration of reality, it is necessary to experience the sea as an image.

5

PASAJEROS A INDIAS

Registers and Biographical Writing as Cultural Techniques
of Subject Constitution (Spain, Sixteenth Century)

A FOUNDING FIGURE OF THE MODERN NATION-STATE?

For year after year throughout the sixteenth century the passenger lists of the Casa de la Contratación in Seville recorded traces of fleeting existences:[1] little bits of information about individuals who entered the stage only to swiftly exit. They appear on these sheets of paper because they were "sending themselves away" (*se despacharon*) from Spain to *las Indias*, the New World (Figure 5-1). Priests, merchants, bookkeepers, trainees, families, single men, wives following their husbands—almost imperceptible traces of obscure lives lacking any *fama*. It requires the melancholy soul of an archivist to take full measure of these registration efforts in all their pedantic monumentality.

These records of the Casa de la Contratación mark the preliminary ending of a long bureaucratic procedure. Since there was no rival agency—be it the Church or any private enterprise—capable of monitoring these migrations, the Spanish state had very early on successfully monopolized "the legitimate means of movement" across the ocean.[2] As a result, the passenger traveling from Seville to the New World was a figure produced with great bureaucratic tenacity. His legal mobility had to be distinguished on the one hand from the dangerous mobility of idle vagrants, beggars, and adventurers (that is, from the notorious *pícaro*), an operation that appears to prefigure one of the foundational distinctions of the modern nation-state, whose "development . . . has depended on effectively distinguishing between citizens/subjects and possible interlopers."[3] On the other hand, the legal passenger was produced by means of excluding all persons of Moorish or Jewish origin. The threshold between

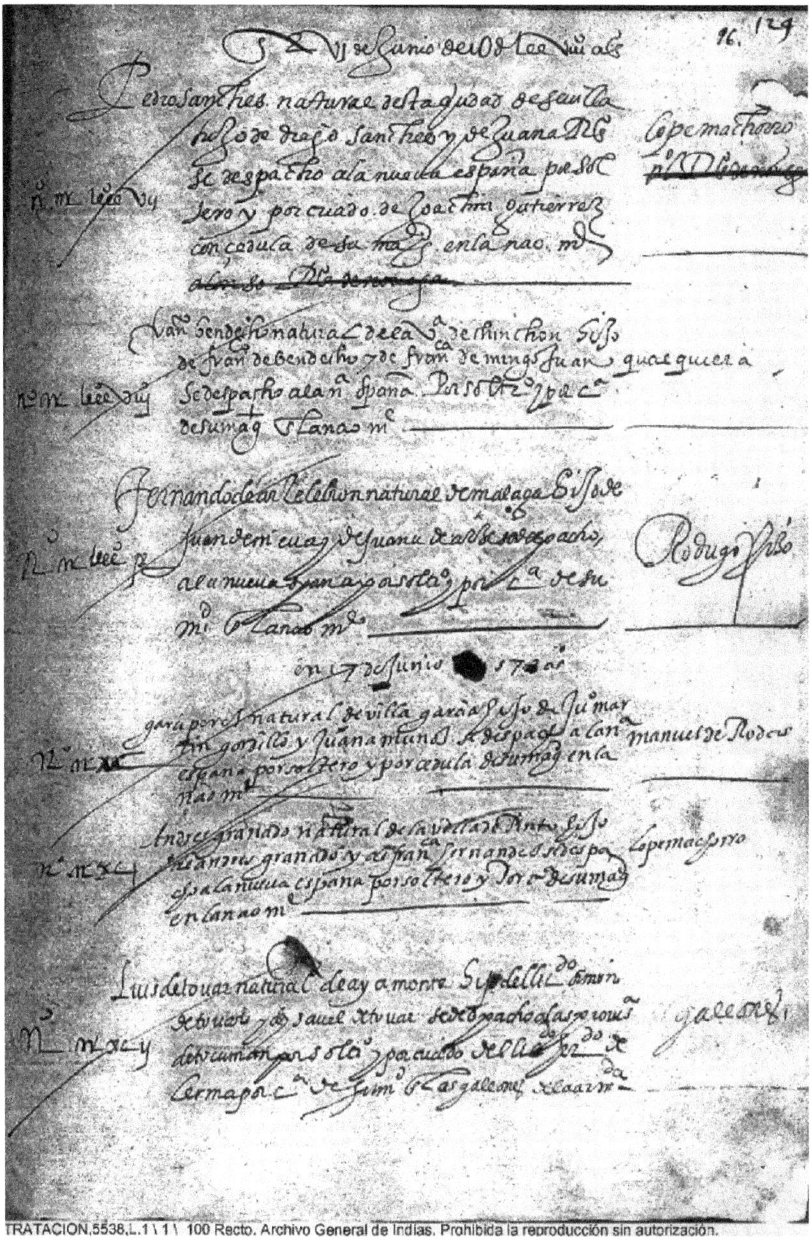

FIGURE 5-1. Casa de la Contratación passenger list, 1553–1556. Archivo General de Indias (AGI), Contratación, 5538, L. 1/1, folio 100r.

land and sea, Spain and *las Indias*, was the place where it became necessary to reproduce the difference between the *cristiano viejo* and the "New Christian"— that is, the *converso* and the *reconciliado*. Both exclusions could only be achieved by fully subjecting the mobility of the passenger to the power of writing as embodied and performed by its agents (the *letrados*) and their institutions. Yet the process of licensing could always be abused as a cover-up for parasitic forms of nonsettled life. As late as the seventeenth century, viceroys in New Spain and officials in the Casa de la Contratación were convinced that the whole process of licensing had the detrimental effect of populating *las Indias* with vagabonds. Discursive practices and administrative techniques were called for to refine this delicate bureaucratic procedure—practices of authentication that turned people that hitherto had lived below the threshold of public discourse into individuals, by making them speak of themselves.

How the sixteenth-century Spanish authorities came to embrace, probe, and register the passengers traveling to the Indies could easily be presented as a prehistory of modern passport controls. Valentin Groebner has in fact offered such a story in his study *Who Are You? Identification, Deception, and Surveillance in Early Modern Europe*.[4] My goal, however, is not to explore the putative origin of migration control, no matter how relevant it may be to the present situation of globalization. The control of migrations in sixteenth-century Spain belongs to the posthistory of the *reconquista* of the Iberian peninsula as well as to the prehistory of the nation-state—as a history of those media and practices that came to constitute the modern subject as an autobiographical animal. We have to rethink the notion that writing and bureaucratization are part of an overall history of modernization and rationalization. Modernity is a dialectic, if not an anachronistic, concept. Before we start to subsume all histories under the one great narrative of the nation-state, we should listen carefully to the noise of the ongoing war beneath the order of the state.[5] Starting in the 1480s, the Reconquista increasingly turned into a war against the invisible Jewish and Moorish elements in Spanish society. In Spain, and especially in Andalusia, the governmental concept of the population took on the shape of an "inner enemy." What makes Spain such a revealing historical example is that it based the construction of the "nation" on the repulsion or conversion of all non-Christian elements. And what makes it an equally revealing example of the premodern state is that it promoted a society in which the subject was characterized by a gap between "the persona" and the body equipped with its cultural signs.

The banks of the Guadalquivir are not only the geographical threshold between the Old World and the New, but also a threshold of description and narration. Nobody crossed into the West without first negotiating this boundary.

From 1530 on, anybody who wanted to travel to the Spanish kingdoms in the New World needed a royal license. Sometimes the process began with a letter to the Casa de la Contratación in Seville, such as the one penned by Catalina de Ribas on July 23, 1585. *"Digo"*—"I say"—she wrote, "that I am the legitimate wife of Thomas de Ribas, who lives in the city of Cartagena in the province of Tierra Firme, and I and my daughter are languishing in great poverty, and I am longing for my husband [*tengo avida de mio matrimonio*]." Her husband, she continued, "is constantly busy in the service of His Majesty in those provinces, and therefore he cannot come to Spain. And by the letters he has written to me and my mother he has compelled me to call upon you, so that I and our daughter might travel over there to live with him."[6]

Catalina could be sure that her request would be heard. Her wish to rejoin her husband complied with the law as well as with a massive campaign aimed at forcing emigrated husbands to have their wives follow them to the Indies.[7] The Casa was authorized to issue licenses for abandoned wives, merchants, and their factors.[8] Everybody else had to write a letter to the king stating their case. If the appeal was successful, the Crown issued a *real cédula*, a royal decree, to the specific petitioner. But the bureaucratic travails were long from over, for in order to turn the license into something useful it was necessary to identify oneself *in corpore* in front of either the *alcalde*, the *corregidor*, or a *juez* in one's hometown. Another petition was required in order to obtain a judicial certification of one's existence, origin, identity, faith, and proper way of life. Furthermore, from 1535 on we find indications that witnesses were examined to testify to the legitimate origin of the prospective passengers. After 1552 the machinery of description grew to its full size as a consequence of a *real cédula* issued by Philip II:

> From now on the Juezes Oficiales will not allow any passenger to pass into whatever part of the Indies . . . if the latter do not bring along and present documents which they have procured in their hometowns and by which they produce evidence whether they are married or not, and about the distinguishing marks and their age and that they are not newly converted to our Holy Catholic faith from either Judaism nor Islam, that they are not *reconciliados* and neither children nor grandchildren of persons who carried the shirt of penance in the public, and neither children nor grandchildren of burned persons or persons who were convicted of heresy . . . together with

a confirmation of the court of the town or village where this information comes from in which it is stated whether the person who gives such information is free or married.[9]

Those who wanted to travel to the new kingdoms overseas had to bear witness to their identity, origin, way of life, scars and birthmarks.

> For a long time ordinary individuality—the everyday individuality of everybody—remained below the threshold of description. To be looked at, observed, described in detail, followed from day to day by an uninterrupted writing was a privilege. The chronicle of a man, the account of his life, his historiography, written as he lived out his life formed parts of the rituals of his power. The disciplinary power reversed this relation, lowered the threshold of describable individuality and made of this description a means of control and a method of domination. It is no longer a monument for future memory, but a document for possible use.[10]

The Casa in Seville is one of the very first sites in early modern Europe where juridical procedures forced hundreds and thousands of simple existences to deliver a written account of their origin, orthodox faith, and orderly life to a representative of the king. Those hitherto untouched by writing are now put on record. What was once a privilege of the few becomes a burden of the many.

Questioning the witnesses served to extract an "individuality" from the silent body. After the witnesses had sworn by the sign of the cross and the name of God to tell the truth, they were examined about the contents of their petition. Usually, the first question was whether and how well the witnesses knew the petitioner, his parents and grandparents, which is why all the witnesses were over sixty years old. The second question concerned the petitioner's *limpieza de sangre* ("purity of blood"). The stereotypical response consisted in the assurance that the petitioner's parents and grandparents had been *cristianos viejos* and of neither Moorish nor Jewish origin. In the third place the judge or *alcalde* (mayor) asked how old the petitioner was (always "un poco mas o menos"), and whether he or she had ever been sentenced by the Santo Oficio. The fourth question addressed marital status. In almost all cases, the witnesses confirmed with great redundancy that the person willing to emigrate was "mozo soltero," a single lad, free and not married.[11] In the fifth place the witness confirmed that the petitioner had never been a member of a religious order (there were special regulations for monks and other clerics). The sixth question concerned the demeanor and character of the petitioner. "He is very quiet and peaceful," one witness informed the judges about Francesco Hernandez, a farmer's son, "and he never rampaged through the villages nor has he been a troublemaker in former times. He is leading a good life and is held in high esteem by everybody and has good manners, and he stopped making trouble and noise and rampaging about."[12]

The seventh and last question addressed the special physical characteristics of the petitioner: body size, complexion, beard color, birthmarks and scars. "This witness knows," one statement reads, "that the said Francesco Hernandez is a man with a small body, who is dark in his face and has a black beard and that he carries a small birthmark under the left eye and a scar above the right eyebrow."[13] "He is a healthy young man," one reads about Alvaro Rodriguez de Mendaña, "whose beard is now starting to grow, and who has a somewhat red face and some freckles on his hand."[14]

In their declarations the witnesses only repeated what the petitioner had already said or written before; the petitioner, in turn, only repeated what the text of the law prescribed; and the scribe put down *en limpio* what the witnesses said. Ultimately, the legal writing of the royal scribe replaced the private writing of the petitioner. But what kind of truth is produced by this ritual? What kind of correspondence (*adaequatio*) serves as a criterion of truth? It is neither the correspondence between the details submitted by the petitioner and the details recalled by the witnesses, nor a correspondence between signifieds. It is rather the repeatability itself that is checked, a repeatability that assures an agreement between spoken and written discourse. The oral utterance has to prove that it is a repetition of the written, and the written words have to prove that they are an anticipation of the oral speech they subsequently repeat. The written discourse, then, must be true because it anticipates the witness's answer to the questions. It is a truth based in the power of literacy over orality. Only when life has been proven to be a repetition of the life contained in documents can it turn into legitimate life. Whatever life may be, it is only in and as writing. By subjecting the unwritten life to the priority of legal writing, the *bios* becomes an element of the political body. Ordinary people become legal persons.

Then it was time to undertake the first step toward America. Leaving behind parents, house, village, hometown, possibly also wife and children, the petitioners headed to Seville, where their first stop was the Casa de la Contratación. Whoever wished to travel to the Indies had to appear in front of the *jueces-oficiales* of the Casa and present all the relevant *informaciónes, fees,* and *testimonios.* The Casa was the Great and Only Lock, the bottleneck between the Old and the New World (Figure 5-2). Everybody and everything bound for the latter had to step into this spotlight to be registered. The judges of the Casa checked the *informaciónes* of the prospective traveler and examined whether the person standing in front of them was in fact the person the documents referred to. Then they gave the order to issue a license to the passenger. With

FIGURE 5-2. Garden of the Consejería de Obras Publicas y Transportes, Casa de la Contratación, Seville. Photograph by the author (2001).

FIGURE 5-3. *Libro registro de pasajeros a Indias* (1588). Reprinted from Gonzáles García et al., *Archivo General de Indias*, 146.

this license the passenger proceeded to the *contaduría* of the Casa. The passenger's name, parents' names, destination, possibly the names of the passenger's spouse and children, profession (in case of clericals or *criados*, servants), and the name of the captain of the ship to which the passenger was allocated were entered into the "Libros de asientos de pasajeros a Indias." Here they finally appear in numerical order, one after the other (Figure 5-3).

FICTITIOUS IDENTITIES

Traces of ordinary lives—witnessed by legal courts, acknowledged by the judge of the Casa. All that had been told and told again, all the testifying, narrating, and examining required by the bureaucratic procedure, gained objective reality by being condensed into a few lines in the "Registro de pasajeros a Indias." The style of the register differs significantly from that of the *pedimientos* and *testimonios*. In the latter everything—name, marital status, origin, age, and so on—was just an assertion, a subjective and in principle dubious claim. "In the city of Ecija there appeared a man in front of His Great Magnificence Señor Alcalde, who said of himself that his name was Diego Ordoñes."[15] This is literature. It contains nothing of which you can say that it is truly so; at best you can state that somebody else said so. The registers, however, speak a different language. They turn assertions into objective reality: What was at first merely said is now a given. There is no *dixo* in the registers and no *sabe*, none of the operators employed by the testimonies to characterize every piece of information as a reported statement. *Pedimientos* and *testimonios* indulge in narration; they are all literature. Registers move beyond narration; they are no longer literature. All the parentheses are excised; what subsequently appears in the registers is no longer a reported claim but acknowledged reality. This *is* Andres Hernandez and he *is* the son of Rodrigo Hernandez and Marie Hernandez; he *is* single and on his way to his aunt in Chile. So it is written.

But is this really no longer literature? There was at least one subject in the Spanish Empire paranoid enough to know better: none other than Philip II himself. If one opens the first volume of the extensive "Libros de asientos de pasajeros," one will find between the cover and the first page a *real cédula* (probably dating from the year 1553). It is addressed to the officials of the Casa de la Contratación and expresses the general suspicion that the stories told to the royal officials are completely bogus.

The Prince
To the officials of the Emperor and King, nuestro señor, who reside in the city of Seville in the Casa de la Contratación de las Indias. We have been

informed that many of the passengers who come to the Casa de la Contratación to deliver the *informaciones* which they produced in their hometowns, as We have ordered with regard to the licenses, present false witnesses [*testigos falsos*] to prove whatever they want, which is the reason why so many who are married state that they are unattached and commit all other kinds of frauds.[16]

The prince has recognized that the whole protracted licensing procedure cannot exclude acts of deception. It is impossible to distinguish truth from fake, reality from ruse, official from fictitious identity. Whoever placed Philip's *real cédula* at the beginning of the "Libros de asientos de pasajeros" thereby gave the royal words the function of a parenthesis—either in an act of ironic whimsy, or with the clear intention to warn the reader of the registers not to trust the information they contain. It is as if the future king had drawn quotation marks around all the entries. Placed at the beginning of the registers, the royal *cédula* becomes a guideline for reading the registers of the passengers: namely, as possible fiction. Thus the aura of the real surrounding the dry statements of the registers represents neither a turn from fiction to fact nor the end of literature. On the contrary, it is an effect of dissimulation, produced by the passengers' *rite de passage* into their new status as legal subjects crossing the threshold between the Old World and the New.

It is therefore the desire of the passengers appearing in front of the representatives of the Great Other to have their existence acknowledged. "I beg Your Mercy to acknowledge that I am what the witnesses will say," implored Alvaro Rodriguez.[17] The erasure of all the parentheses that marked the information in the *interrogatorios* as subjective speech serves to validate the existence of a person as the real-life referent to the *informaciónes* provided. After all, the existence of ordinary people is a contingent matter that can be derived neither logically nor ontologically. There is no memory that could testify to their existence, no chronicle of marvelous ancestral deeds, no genealogies, no residences named after them. All of their existence is contained in the handful of words that tell who they are. Their very being therefore depends exclusively on being acknowledged by the law. This being, the existence of a referent, *mise en scène* by registers, is placed between quotation marks by the prince. The very register that turns individuals into subjects of the royal technologies of writing endangers the social body of the state because it dissimulates the fiction it contains. Ultimately there is something worse than missing entries in the passengers' registers: namely, entries that are made up of fictitious information. The realism of the "Libros de asientos de pasajeros a Indias" is deceptive. The prince is haunted by the idea that the people listed in the registers could be somebody else. They all could be imposters, con men, cunning *pícaros*.

Despite the well-known connection between *picarescas* and *conversos* in post-Reconquista Spain, the suspicions of Philip II were not in the first place directed at the Christian purity of the Spanish population of the Indies. The principal concern of the Spanish crown was not whether those persons attracted by a life in the West Indies were "nuevamente convertido de Judio o Moro" ("recently converted Jews or Moors"), but whether they were *casado* or *soltero*, married or single. The main goal of king and council was to prevent married men from deterritorializing themselves in the new American colonies. The suspicion that many of the passengers who claimed to be *mozo solteros* (single young men) were in truth married *paterfamilias* reveals the degree to which the Spanish administration and the viceroys in America were less concerned with securing the Christian bloodlines of the population than with the possibility that the lure of making a fortune overseas might be stronger than the bonds of the sacrament of marriage, stronger than love for the wife and children abandoned in the Old World. The specter of thousands of married men leaving their crafts, wives, and children with the promise to return after only a couple of years, heavily laden with gold and silver, haunts the sixteenth-century legislation for the licensing procedures for passengers. The colonial *conquista* threatened to depopulate the Old and populate the New World with vagabonds and adventurers. The Indies were not threatened by dubious converts, but by men who had abandoned their families and professions to enjoy a parasitic existence. As early as 1544, a *real cédula* states "that of these unmarried men none cares either to procreate or to plant or to seed or to build or to raise something or to do anything of the kind a good settler would do."[18]

Stowaways or passengers who had obtained their license by fraud, wrote the viceroy of New Spain to Philip II, were crowding the kingdom of New Spain with so many vagabonding idlers (*bagamundos holgazanes*) that they would strangle the country. The New World was a paradise for idlers, a realm of lazybones, gamblers, and loose women. Of the hundreds of passengers who arrived from Seville every year, none wished to work, complained the viceroy. "None wishes to enter somebody's service and to work, and the streets and places are crowded with idle and futile women and vagabonding spoilt men."[19]

THE PARASITIC ECONOMY OF THE FICTITIOUS

Spain has contributed significantly to the European discourse on the figure of the idler: As the home of belligerent adventurers lusting after gold, Spain is the land of idlers par excellence. The conquistador is, as it were, the most triumphant incarnation of the idler, though this cannot hide the fact that the *pícaro*, the sly beggar, is his poor relative.

The fear that the parasitic economy of deterritorialized passengers could lead to the ruin of the Spanish overseas empire arose in connection with a loss of meaning affecting poverty and vagrancy. This loss of meaning, which gained momentum in the late Middle Ages and was especially pronounced in Spain following the Reconquista, had a decisive impact on medieval culture and helped shape the face of modern Europe. Indeed, both poverty and vagrancy are particularly revealing when it comes to studying the ambivalent effects that emerge when cultural techniques of identity and disciplining start to erode and displace a medieval culture based on symbolizations.

Towards the end of the fifteenth century the medieval "culture of poverty" became increasingly ambiguous. In contrast to the pagan cultures of Greece and Rome, which viewed poverty as misfortune,[20] Christianity had insisted that the poor play an indispensable part in the *oeconomia sacra*. Evangelists and Church fathers praised poverty (*paupertas*) as a sign of humility (*humilitas*), but early Christian literature does not only extol voluntariness as a fundamental component of poverty, it also praises the external signs: threadbare clothes, lack of possessions, a life without a home.[21] Homelessness occasionally entailed that paupers were seen as strangers: To be poor meant to be on the road. Because Christ's earthly existence had sanctified poverty,[22] the poor were identified with Christ. In line with the Church's penchant for corporeal imagery, which conceived of the body of Christ as a real metaphor of Christian community, the poor came to be seen as the "suffering limbs" of Christ and as representatives of Christ on earth.[23] In the Christian hierarchy of offices, the poor occupy the highest position: Pierre de Blois refers to them as "vicarius Christi,"[24] a title reserved for medieval rulers before a decree of Pope Innocent III from around 1200 turned it into the official papal title.[25] The poor are thought to be privileged intercessors before God. Their prayers result in particular blessings for those who chose them as their advocate; hence the custom of having a poor person hold the child over the baptismal font. To be poor is a kind of ad hoc office that can be held by anybody displaying the external signs of poverty.

In line with this Christian economy of redemption, the giving of alms complements the office of the poor and endows it with meaning. Alms are not only a means of expiating particular sins; for the rich they are the only means of attaining eternal salvation. The poor are therefore a necessary element of God's design. The *Life of St. Eligius* gets to the heart of the matter: "God could have made all men rich, but He wanted there to be poor people in this world, that the rich might be able to redeem their sins."[26] To extend hospitality to the itinerant poor, the vagabonds, is to take in Christ. "Let the greatest care be taken, especially in the reception of the poor and travelers," runs one of the rules of Saint Benedict, "because Christ is received more specially in them."

Towards the end of the Middle Ages the Christian doctrine that invested poverty with a well-defined function within the *oeconomia sacra* encountered a formidable challenge. It was increasingly forced to compete with a discourse that described the poor and the vagrant as a threat to society, a social evil.[27] The central distinction of this new discourse was that between *mendicantes validi* and *mendicantes invalidi*, that is, between true and false beggars. Poverty became ambiguous; and the more dubious it became, the more the preordained office of the poor in the salvational economy was eroded. The distinction between *mendicantes validi* and *mendicantes invalidi* derived from a notion of poverty that conceived of the latter exclusively in terms of one's relation to work. "True" poverty is the result of the inability to work—be it due to age or physical infirmity. But those who beg despite being able to work are bums and idlers who merely fake poverty. The thirteenth-century Siete Partidas, a vernacular statutory code introduced during the reign of Alfonso the Wise of Castile, distinguished between *validos medicantes* to whom hospitality must be extended and impostors who begged without need. The latter were to receive no alms and be chased out of the country if they were healthy in body.[28]

But in contrast to medieval revisions of indiscriminate neighborly love, sixteenth-century discourses and practices aimed at mendicants were less concerned with the necessity to distinguish between "true" and "false" beggars than with the problem of how to tell them apart in the first place. An ontological issue turns into a practical problem. In the German-speaking countries of Europe, the *Liber vagatorum* revealed the deceptive scheming of fake beggars. Luther contributed an introduction to the eighth edition.[29] The Italian *Speculum cerretanorum* had appeared earlier. The early fifteenth-century *Basler Betrügnisse* ("Frauds of Basel") listed twenty-six different types of fake beggars; the *Liber vagatorum* extended the list to forty-one.

The inevitable result was a profound disruption of the medieval economy of salvation according to which the poor act as representatives of Christ capable of opening the gates of heaven for those who had given them alms. Salvation was no longer guaranteed, if charity had been wasted on those who were not qualified to act as advocates. In such a case, the performative act of giving is as null and void as a marriage blessed by a fake priest. Referring to Matthew 19:29, Pérez de Herrera noted: "God will not return our alms a hundredfold because he never received them in the first place."[30] The pauper, formerly a potent symbolic figure, has turned into a mask of nonmeaning. The economy of salvation, in which the poor had bestowed a meaning upon the acts and lives of the rich, is eroded. In its midst there now resides a parasitic economy that redefines the relationship between the meaningful and that which threatens meaning, the "true" and the "false." The idler is a freeloader, a parasite of the symbol. That is to say: He lives off the symbol, but in doing so he renders it useless. He deprives society of meaning as well as money.

INFINITE PROGRESSES OF CERTIFICATION

The suspicion that nobody is what he claims to be is not only aimed at the ordinary people who have become subjects of a legal passage, or at the *conversos*; it is also directed at the *letrados*, that is, the judges, mayors, and scribes. The testimony of the witness is frequently accompanied by a "fee de Juez" with which the mayor certifies the identity and truthfulness of the witness. This certificate, in turn, is accompanied by another certificate from the hand of another royal scribe attesting to the fact that the mayor is in fact the mayor and the scribe is, really, a royal scribe (Figure 5-4):

> I, Antonio Velez, public scribe of His Royal Majesty . . . certify and give true evidence to all lords who may see and read the document on hand, that the said Juan de Ortega, who has signed this Informacion and this certificate, is a scribe of His Royal Majesty and one from the number of the said city, and that he is a scribe who is reliable in his office and law-abiding. . . . And likewise I certify that the said Alvaro de Soto, who has signed this, is the regular mayor of the said city of Almazan, and I recognize that this is his signature because there are a lot of his signatures in my documents that are like the signature here [*conozco ser su firma por q(ue) muchas firmas tengo yo en mis scripturas que son como la que aqui firmo*]. . . .
>
> [Scribe's mark] In testimony of this truth [*En este testimonio de verdad*], Ant[oni]o velez, *scribano*[31]

This particular example illustrates that identity can only be attested to if the sign by which it is recognized has already been copied. By testifying that there are many copies or repetitions of the signature of the mayor, Antonio Velez certifies the identity of the actual signature. The copy always precedes the original. The deconstruction of the occidental concept of authenticity is an everyday practice of bureaucracy. Inevitably such a strategy of certification leads to an infinite regress: The certification of the scribe Velez is followed by a certification from the hand of another royal scribe named Alonso Perez de Palma, who testifies that the alcalde is in fact the alcalde and Juan de Ortega is in fact a scribe in Almazan.[32] This is followed by a series of "frames." The text of the certification is framed by the scribe's mark (which is something between a signature and a seal); the mark, in turn, is framed by the signature, which is framed by squiggles that fill the empty spaces of the paper up to the margin in order to prevent later additions. Finally, everything is framed by the words "es

FIGURE 5-4. Certificate of authenticity ("fee e verdadero testimonio") by the scribes Antonio Velez and Alonso Perez de Palma from Almazan, 1563. AGI, Contratación, 5221, N. 1, R. 3/1, folio 6v.

bastante"—"it's enough."[33] There is no last frame. The infinite regress of frames can only be broken off by an arbitrary act.

All passengers could be disguised idlers. But it is not the passengers who upset Spanish reality by importing a world of deception and as-if. The protracted discursive practices of examination, description, and validation are to blame for populating the Indies with phantoms. At the end of the bureaucratic process obsessed with identifying and legitimizing the passengers, the resulting protocols appear to the suspicious prince as an abyss in which language itself disappears. The discursive practices designed to acknowledge ordinary people as legitimate passengers are based in their entirety on the "non-serious" or "parasitic" use of language John Austin (1962) wanted to exclude from his theory of speech-acts: They depend on citationality.[34] The petitioner's *escritura de pedimiento* quotes the wording of the royal law, the witnesses quote the questionnaire of the petitioner, the royal scribe quotes the testimony of the witnesses, and the register entry of the Contador in the Casa quotes the passenger's *información* previously acknowledged by the *jueces* of the Casa. In the same way the oral testimonies of the witnesses confirm the truth of the passenger's story simply by repeating the facts contained in the documents. Deception and mendacity are not an "external place of perdition" that could be avoided; they are, "on the contrary, [the] internal and positive condition of possibility" for the bureaucratic construction of reality. [35] The fake is not the perversion of the fact, or the exception of the vain and frivolous from the rule of the serious and referential—no, the fake is inside the fact.

6

(NOT) IN PLACE

The Grid, or, Cultural Techniques of Ruling Spaces

THE GRID AS CULTURAL TECHNIQUE

Xenophon's *Oeconomicus* introduces *taxis* as a fundamental cultural technique of the economic domain. *Taxis* refers to an order of things in which each and every object is located in a fixed place where it can be found. Humans, however, differ from things. "When you are searching for a person," Xenophon cautions, "you often fail to find him, though he may be searching for you himself."[1] Humans defy the fundamental rules of economy because for them "no place of meeting has been fixed."[2]

This distinction between retrievable things and untraceable humans points to the fundamental divide that separates the Greeks from modern subjects. Modernity is characterized by the invention of a *taxis* technique capable of also turning humans into retrievable objects. This modern *taxis* is implemented by means of a new cultural technique which takes into account that something may be missing from its place. In other words, it encompasses the notion of an empty space. The technique in question is the grid or lattice. Its salient feature is its ability to merge operations geared toward representing humans and things with those of governance. As Deleuze noted in his study of Foucault, between the sixteenth and eighteenth century, grid-shaped control becomes the universal practice that constitutes the basis of modern disciplinary societies.[3]

The ontological effect of the grid is the modern concept of place and being-in-one's-place based on the media-theoretical distinction between data and addresses. In other words, it presupposes the ability to write absence, that is, to deal equally efficiently with both occupied and empty spaces. This concept of place is thus inextricably tied to the notion of order. In return, it is impossible to conceive of this modern concept of order without the new understanding of place.

The universality of this concept of order is apparent in the way in which it bears on the interaction between imaging technologies and mathematical, topographical, geographical, and governmental knowledge. It is this interaction that turns the grid into a cultural technique. But what does this imply? As a cultural technique, the grid has a triple function. First, it is an imaging technology that by means of a given algorithm enables us to project a three-dimensional world onto a two-dimensional plane. That is, it is a type of representation that posits an antecedent geometrical space in which objects are located and that submits the representation of objects to a theory of subjective vision. Second, the grid is a general diagrammatic procedure that uses specific addresses to store data that can be implemented in the real as well as in the symbolic (grids may be two- or three-dimensional, or 2D/3D hybrids). Third, the grid serves to constitute a world of objects imagined by a subject. To speak with Heidegger, it is a *Gestell* or "enframing" aimed at the availability and controllability of whatever is thus conceived; it addresses and symbolically manipulates things that have been transformed into data. The grid, in short, is a medium that operationalizes deixis. It allows us to link deictic procedures with chains of symbolic operations that have effects in the real. Hence the grid is not only part of a history of representation, or of a history of procedures facilitating the efficient manipulation of data, but also of "a history of the different modes by which, in our culture, human beings are made into subjects."[4]

As important as the distinction between centrifugal and centripetal grids may be,[5] for the purpose of analyzing the grid as a cultural technique it is more relevant to distinguish between representational, topographic, cartographic, speculative, and three-dimensional (total) grids. This division, in turn, suggests a more fundamental question: Can the expansion of Western culture from the sixteenth to the twentieth century be described in terms of a growing totalitarianism of the grid?

REPRESENTATIONAL GRIDS

The fact that the grid effectively merges representation and operation is already apparent in Leon Battista Alberti's 1435 treatise *De pictura* (*On Painting*), which deals with grids as part of an imaging theory. Alberti's famous *velum* (veil) is a *perspectiva naturalis* technology designed to circumscribe objects (*circumscriptio*). Together with the window, which serves as metaphor for the mathematical construction of paintings, the veil is a medium for their technical construction. Alberti also refers to it as an intersection or *intercisio*, thereby linking it to his definition of the image as an intersection of the visual pyramid. The veil

> [is] woven of very thin threads and loosely intertwined, dyed with any color, subdivided with thicker threads according to parallel partitions, in as many squares as you like, and held stretched by a frame, which [veil] I place, indeed, between the object to be represented and the eye, so that the visual pyramid penetrates through the thinness of the veil.[6]

Alberti's veil is the basis for all technological imaging procedures, until the twentieth century, that employ reprographic techniques such as hole patterns and halftone. Since the late nineteenth century, industrial graphics has used coded perforation patterns to resolve and make transmittable a template's halftones. Thus the veil survives in today's screen printing technologies.

But already in the seventeenth century halftone techniques had been linked to a media-theoretical modeling of neuronal signal processing. The mezzotint print technology invented in 1642, which consists in wiping off the surface of a burnished copperplate,[7] corresponded to Descartes's physiological theory of the processing of optical perceptual data by their analytical decomposition into hole patterns that are engraved into the brain. But technical images do not have to wait for the arrival of copperplate engravings or the idea of neuronal signal processing by means of hole patterns; they precede the age of their technological reproducibility. Technical images, in other words, are no exclusive hallmark of modernity. The textile image was always already a technological image because it was produced by the mechanical distribution of warps and wefts.[8] It comes as no surprise that Alberti resorts to weaving and textile images when discussing the intricacies of central perspective. Indeed, at the beginning of the fifteenth century it was quite common in Europe—for instance, at the court of the Duc de Berry—to consider tapestries to have higher value than pictures. Alberti is so firmly rooted in the textile paradigm that his claim to have produced a scientific—that is, mathematical—treatise is constantly thwarted by explanations that harken back to the art of weaving. One example of many is his attempt to give a textile spin to Euclid's definition of a surface: "If more lines stick together like close threads in a cloth, they will make a surface."[9]

According to Alberti, the great asset of the veil is "that it always presents the same surfaces unchanged," because it fixes the apex of the visual pyramid. "And so, the veil will guarantee this not negligible advantage which I have spoken of: that the object always stays the same with respect to the view [ut res semper eadem e conspectu persistat]."[10]

To grasp the ontological implications of Alberti's grid technique it is necessary to emphasize its connection to the categories of place or *locus* in his treatise:

> Since painting, in fact, aspires to represent the objects seen, let us note in what way they themselves come to sight. First of all, when we watch an [object], we certainly see that there is something that occupies a place.[11]

Something real (*res*) is something that occupies a space, that is in its place. It is crucial to be mindful of the connotations that the term *locus* (Greek *topos*) possesses in both rhetorics and *ars memoriae*. As Alberti would have it, only that which occupies its place is a representable object. Whatever lacks an identifiable place—such as the old gold ground or the halos that Alberti disdains—cannot and should not be depicted.[12]

Alberti's grid is an ordered space: a space in which aesthetic, ontological, and diagrammatic orders exercise their power over the existence and appearance of objects. It is a space that pays tribute to the power of cultural techniques to assign things and figures their own space. Hubert Damisch defined this "data space" as the "paradigmatic dimension" of the *costruzione legitima*:

> To each figure its place: at each point on the underlying checkerboard, if not within each of its squares, one figure and only one, among all those that are possible, can be situated.[13]

Damisch makes clear that both the grid and the checkerboard of the *perpectiva artificialis* are structures in the Saussurean sense. The representation of objects in pictorial space implies their substitutability, which in turn reveals the analogy between Alberti's perspectival space and the place-value system of Indo-Arabic numerals. The fact that the grid precedes the object located therein (which implies both the possibility of addressing an empty place and the contingency of whatever object happens to be situated there) is equivalent to the semiotics of zero.[14] Brunelleschi as well as Alberti were members of that particular social class that in the *trecento* first absorbed and circulated knowledge of Indo-Arabic numerals. To each figure its own piece of space, to each numeral its place—in Germany, incidentally, these numerals were still known in the seventeenth century as *figurae*.

It is for this structural reason that digitization is able to retire the *velum*. Once you have two moveable scales and a sighting mechanism such as a cross-staff or a quadrant, a veil is superfluous (Figure 6-1).

FIGURE 6-1. Achieving perspective without the *velum*. Notice the *adlatus* on the floor, holding the grid in his hands, on which he enters the coordinates the artist-engineer is telling him. Engraving, reprinted from Jacopo Barozzi da Vignola, *Le due regole della prospettiva pratica* (Rome: Stamparia Camerale, 1611).

THE CARTOGRAPHIC GRID

Cartographers have used the grid as a technique for ordering space since antiquity. Claudius Ptolemy was the first to wrestle with the problem of projecting a spherical surface onto a plane surface; as well, he pioneered the method of dividing a surface into a lattice of latitudes and longitudes. In the second century he authored a treatise on cartography. His *Geography* was in all likelihood expanded by Byzantine scholars, and upon reaching Italy was translated into Latin in 1406. Ptolemy probably only wrote book 1, the beginning of book 2, and chapters 3 through 28 of the final book 8. The latter contains the coordinates of roughly three hundred cities based on time measurements. The longitude is determined by the temporal distance from the Alexandrian meridian, with one hour corresponding to fifteen degrees of longitude. The latitude is determined by the length of the longest day. The greater the distance from the equator, the longer the longest summer day in the Northern Hemisphere.[15] Based on information provided by Ptolemy and the Byzantine additions, European publishers added maps to their editions.

The importance of the strict Euclidean ratio of point, line, and surface is already apparent in Edward Wright's map of 1599, which uses an improved Mercator projection to depict longitudes as parallels.[16] Latitudes and longitudes impose themselves on the medieval system of rhumbs. The grid thus turns into a diagram,[17] enabling the depiction of temporal sequences in addition to spatial orders. Once we read the synchronic segments diachronically, time appears as a function of space.

Ever since the arrival of matrix screens in the early 1970s, the addressing of points by means of rows and columns has turned into a universal imaging technique. While the imaging technique of vector graphics corresponds to the navigational technique of medieval portolan charts based on rhumb lines (a point is defined by its angle to and distance from the *origo*), the matrix screen corresponds to navigation by means of latitude and longitude. Unlike vector graphics that only store the beginning and end of a line, the matrix screen must take account of every single point on the line. The advantage of the latter is the addressability of every single element on the screen, because the screen memory delegates exactly one unit of storage space to each point.

THE TOPOGRAPHIC GRID (SOUTH AMERICA)

One of the effects of the representational technique known as central perspective is that the identity of objects becomes a function of their being in a particular place. The navigational technique employing latitude and longitude, in turn, enables us to head for any point in space by means of addresses that precede all stored data. The overall result is a common paradigm of image construction and early modern colonial governmentality that far exceeds the boundaries of art. To put persons or things in their place, to objectify and subjugate them, are procedures Heidegger already detected in the term *repraesentare*.[18] Turning to the colonial topographies of Spanish-American settlements, we can see that they are superimposed on a grid of very different provenance: the checkerboard design of urban planning associated with Hippodamus of Miletus. The representational grid of the Renaissance decodes and recodes the early grid of colonial topography devised by antiquity.

What results from this superimposition of the representational on the urban grid? Colonialism unleashes and mobilizes the utopian social potential contained in the grid-shaped heterotopias of Latin America. Three aspects are of particular importance: a) the possibility of registering the absent; b) the distinction between data and addresses; and c) the potentially infinite extension in time and space. The latter marks a decisive difference between the grid of Spanish colonial topography and Alberti's veil: The contained *velum* is a figure, whereas the centrifugal orthogonal net of colonial settlements is not.[19] The grid is thus located at several junctions: It straddles the boundary between antiquity and the modern age, and it marks the transition from the political to the economic (or governmental) as well as from symbolically organized space to graphically coded surface.

The origin of Spanish-American checkerboard cities has been the subject of lengthy debates. In all likelihood they arose from medieval and early modern attempts to adapt the construction of Roman military camps. A medieval Spanish treatise on urban planning containing, among other items, the *Regiment de Princeps* by the Catalan Franciscan Francesc Eiximenis (app. 1340–1409), describes the ideal grid-shaped city.[20]

The Roman *castrum* reflects the practice of centuriation. Centuriation is the division of the land into square units called centuries, carried out by the *agrimensores*, the Roman land surveyors. The more Rome came to dominate Italy, the greater was the need to divide up public domains and found new colonies.[21] The only surviving official Roman survey maps are the cadasters of Arausio (Orange, Vaucluse); in addition, there is an extant collection of Roman land surveying records known as the *Corpus Agrimensorum Romanorum* that contains treatises dating as far back as the first century B.C.E. The principal Roman measuring instrument was the *groma*, used to lay out straight lines, right angles, and squares. The usual procedure was for the land surveyor to divide up the land, draw lots for the landholdings, and lead the settlers to their fields. He was also responsible for making a map (*forma*) and compiling a register. "Between each pair of centuries was a *limes*, literally 'balk,' for which appropriate width was provided; in one direction each of these constituted a *kardo*, and at right angles to it was a *decumanus*."[22] A grid was organized and its sections named by reference to two main roads or axes, the *decumanus maximus* and the *kardo maximus*. This is how the grid appears in the *Corpus Agrimensorum* (Figures 6-2 and 6-3).

Erwin Walter Palm has pointed out that when it came to cadasters, colonies, and *castra*, Spain adopted features of the Roman system even before 1492. What was subsequently transferred to the New World followed in the footsteps of the Reconquista.

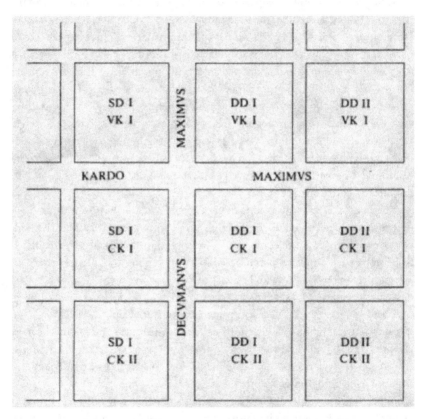

FIGURE 6-2. Centuriation. SD = sinistra decumani (left of the decumanus maximus), DD = dextra decumani (right of the decumanus maximus), VK = ultra kardinem (beyond the kardo maximus), CK = citra kardinem (this side of the kardo maximus). These abbreviations were engraved into the boundary stones that marked off the areas. Reprinted from Harley and Woodward, eds., *Cartography in Prehistoric, Ancient, and Medieval Europe and the Mediterranean*, 213.

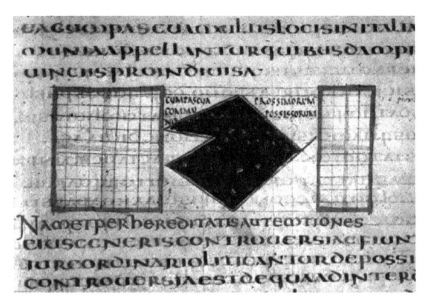

FIGURE 6-3. Illustration of uncultivated pasture land between two centuriated colonies. Miniature from the *Corpus Agrimensorum*. Biblioteca Apostolica Vaticana. Reprinted from Centre de création industrielle and Centre Georges Pompidou, *Cartes et figures de la terre*, 403.

However, there is a conspicuous difference between the shape of the Roman *castrum* and the planned colonial heterotopias: While the former is contained within a square with the four gates located at the ends of the *decumanus maximus* and the *kardo maximus,* the latter call for an infinite expansion. The gridiron that appears on the plans of Lima, Santiago de Léon de Caracas, or San Juan de la Frontera recalls the Greek cities designed by Hippodamus of Miletus. Hippodamus, a contemporary of Herodotus and graduate of the famous Milesian school, lived during the fifth century B.C.E. Aristotle records that he devised the gridiron plan for the port town of Piraeus. He is also credited with the checkerboard layout of Rhodes and of Miletus itself, destroyed by the Persians in 494 B.C.E. (see Figure 6-4). The Hippodamian grid consists of regular squares created by streets intersecting each other at right angles, evidently a concrete realization of that particular type of reason that characterized the Milesian school, which identified urban order with political order.[23] However, archaeologists have long known that the Hippodamian plan, too, had its predecessors, be they the Greek colonial settlements of the seventh and eighth century B.C.E. or the hypothetical Etruscan city plan, which may be the heritage of an Italianate tradition reaching far back into the ages preceding the

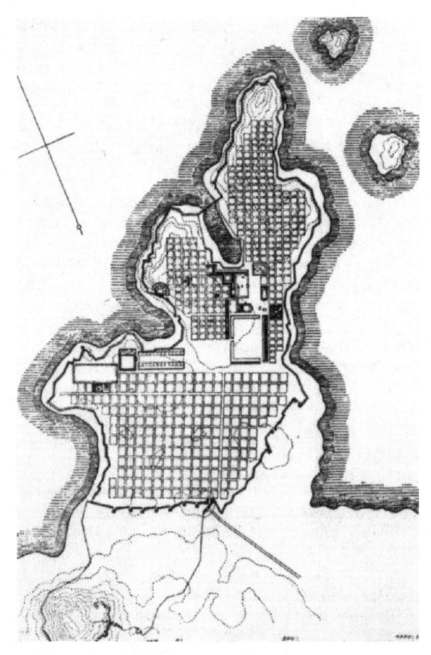

FIGURE 6-4. Plan of Miletus, attributed to Hippodamus. The city of Miletus was destroyed by the Persians in 494 B.C.E. and rebuilt after 479 or 466 B.C.E. Published in Armin von Gerkan, *Griechische Städteanlagen* (1924); reprinted from *Cartes et figures de la terre*, 32.

Indo-European migration. But no matter what: According to Aristotle, it was Hippodamus "who invented the division of cities into precincts and who laid out the street-plan of the Piraeus."[24] However, Aristotle saw Hippodamus less as a pioneer of a new kind of urbanism than as the inventor of a new way of segregating the population into three parts, one of skilled workers, one of farmers, "and [a] third to bear arms and secure defense."[25] Damisch notes that the concept underlying the gridiron pattern devoid of a center is that of "nemesis." In addition to the idea of vengeful fate, nemesis also refers to the notion of distributive law and of giving what is due, that is, to the correlated idea of an expansion in need of boundaries.[26]

In Miletus and Piraeus, the construction of surrounding walls, including the positioning of the gates, depended on geographical features and military considerations rather than on planning. This also applies to the foundation of sixteenth-century Middle and South American cities. Lima was seen as the most typical Latin American city (Figure 6-5). The Jesuit missionary and historian Bernabé Cobo (1582–1657) wrote of its founding in 1535:

> In order to found the city the governor first completed a drawing with streets and city blocks, whereon he noted who was to be assigned which plot by putting down their names; and doing this regardless of the number of inhabitants [vecinos] present at the time of foundation of the city (there were only 69) but with a view towards the size it promised to attain, a space large enough for 117 blocks was laid out. . . . Each had a front length of 450 feet, the settlement was to stretch 13 blocks in width and 9 in breadth, separated by streets, and ropes were used to assure that each street was 40 feet in breadth.[27]

Let us note two key points: First, the plan described by Cobo is at one and the same time plan, register, and cadaster. Second, the city was not planned and built on the basis of the actual number of settlers, or as a means of distributing property, but with a settlement fantasy in mind. This fantasy is enabled and sustained by the possibility of writing empty spaces, that is, the ability to literally reserve a space for the unknown. This, in turn, presupposes the separation of data and addresses. Persons (be they public or private) are turned into data that can be stored for subsequent retrieval by the correct addresses that logically and temporally precede them. The Latin American heterotopia is thus the first concrete realization of the storage model we know today as *working memory*. The 1562 charter of San Juan de la Frontera in the Cuyo region of Argentina reveals the future orientation alluded to by Cobo (Figure 6-6). The suggested continuation of the grid refers to its virtual boundlessness, and the multiplication of the squares is complemented by their internal divisibility.

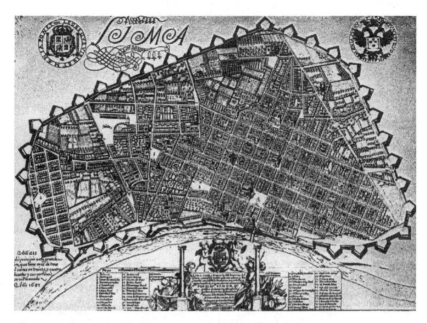

FIGURE 6-5. Map of Lima, Peru, 1687. Reprinted from Hardoy, *Cartografía urbana colonial de América Latina y el Caribe*, 146.

The planimetric settlement plan of Teutenango in Mexico from 1582 appears to have effectively fused scaled paper and *trazado a cordel* (Figure 6-7). The plots turn into an inscription surface, a sheet of scaled paper. The act of *Landnahme*, the taking of the land—in this colonial context, "land grab" would be a more appropriate translation—coincides with graphic operations on the paper surface. Political and diagrammatic space are one.

The *trazados* of the Latin American cities precisely conform to the double meaning of Spanish *padròn*: chart and register. Nothing demonstrates this more impressively than the city plan of Buenos Aires of 1583. It looks like a register, but it is a city plan—a city plan that is both register and cadaster (Figure 6-8). The *nomos* of the earth and the *nomos* of bureaucracy coincide. To live in one of the newly founded Latin American cities amounts to being registered in a grid in which, to quote Alberti, *omnia in locis suis disposita*—everything is assigned its own place. Santo Domingo, Mexico City, Lima, or Buenos Aires are at one and the same time topographic *loci* where people live and memorial *loci* in a storage medium. Cities are both physical space and technological memory.

The letters and memoranda penned by the Franciscan friar Gerónimo de Mendieta to the president of the Consejo de Indias shed light on the ontological status of the these urban data spaces. Mendieta's proposals, aimed at improving

FIGURE 6-6. Foundation charter of the city of San Juan de la Frontera (Argentina), 1562. Reprinted from Gonzáles García et al., *Archivo General de Indias*, 204.

the economic and social conditions in the colonies, revolve around two governmental techniques: the introduction of registers and the foundation of new settlements (grid-shaped, naturally). The bill he drafted concerning the introduction of population registers in the Spanish colonies[28] contains a highly revealing ambiguous formulation: The official in charge of the registers is to "place every single one in his place."[29] What did Mendieta have in mind with this phrase? Is the official to place "every single one in his place" in the register, or at their actual place of residence? The place in the register and the place of residence are made to overlap. How is one to distinguish the symbolic place from the real? This blurring of boundaries is no coincidence. It is, quite simply,

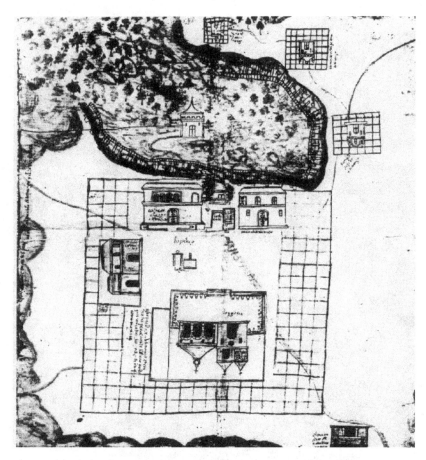

FIGURE 6-7. Map of Teutenango, Mexico, 1582. Reprinted from Guidoni and Guidoni, *Storia dell'urbanistica: Il cinquecento,* 353.

the situation—a situation determined by the culture-technical operationalization of deixis.

This characteristic indistinguishability is also present in Mendieta's settlement projects. "[P]ara poner en asiento los muchos españoles que andan vagueando";[30] the Spanish are to be collected in *pueblos formados,* that is, in grid-shaped cities. *Poner en asiento* means both to settle and to register somebody. Thus Spanish colonialism was able to extract from the law guiding Alberti's constructions—namely, that the being of beings is constituted by their ability to be represented—a disciplinary and governmental dimension by applying the representational properties of Alberti's veil to the Hippodamic checkerboard. The result is, in Angel Rama's memorable phrase, the *lettered*

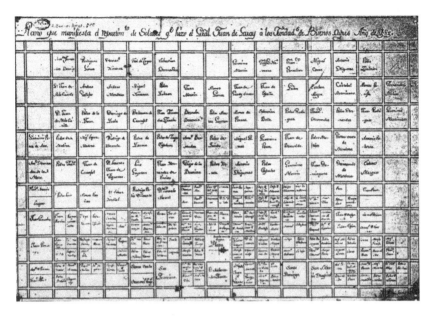

FIGURE 6-8. Plan of Buenos Aires, 1583. Reprinted from Hardoy, *Cartografía urbana colonial*, 67.

city.[31] Reality always already assumes the shape of writing; it becomes a re-peated, quoted reality. By putting individuals into their place in the *padrón* (that is, in the symbolic), they are assigned definitive coordinates in the real. Those who once were lost are now in their place.

THE SPECULATIVE GRID (UNITED STATES)

Two hundred years later: The U.S. Continental Congress passes the Land Or-dinance of 1785. Claims to the territory west of the Alleghenies by "landed"' states such as Virginia in the south and Connecticut in the north as well as the ensuing dispute between "landed" and "landless" states compelled Congress to insist on cessions by the former and the creation of a public domain (the North-west Territories). In the eyes of the landless states, their landed counterparts, which laid claim to the entire region north and west of the Ohio River as well as to the territory of the future state of Kentucky, threatened to dominate the confederation. More importantly, the United States was under pressure to pay off the debt incurred during the Revolutionary War. Deprived of taxing author-ity, Congress viewed the survey and partitioning of the Northwest Territories as a welcome opportunity to profit from land sales for the purpose of servic-ing the debt.[32] Unlike the sixteenth-century South American grid, then, its late eighteenth-century North American counterpart was less a governmental technique than a scheme aimed at capitalizing on federal land. By means of the grid-shaped survey of the territories ceded by the landed states, the United States acquired territory, a public domain to be auctioned off in standardized plots at set prices. Although the rectangular survey prescribed by the Land Ordinance of 1785 only concerned territories between the Appalachians and the Mississippi, it became the model for the subsequent appropriation and colonization of the entire continent. Congress was confronted with a situation virtually unprecedented in history: It was empowered to "make the law governing the survey and distribution of a vast territory before it was occupied."[33]

Following a proposal put forward by Thomas Jefferson, Congress opted for a rectangular survey of straight lines and right angles, such as had been favored by the New England states in the earlier stage of colonization.[34] It basically consisted of projecting a township lattice of latitudes and longitudes onto the territories west of the Ohio. Initially the survey was to limit itself to a strip of land forty miles wide, located to the west of Pennsylvania and extending north from the Ohio into lands still held by Native Americans. Led by Thomas Hutchins, the surveyors divided the territory into townships of six square miles each "by lines running due north and south, and others crossing these at right angles, as near as may be."[35] Each range began at the Ohio with number 1 and was then numbered from south to north, while the seven ranges themselves were numbered from east to west (Figure 6-9).

The township plats were divided into "lots of one mile square or 640 acres,"[36] and were numbered from 1 to 36. Surveyors were instructed to keep a watchful eye on variations of the magnetic needle. As if transferring Simon Stevins's maritime navigational technique to the western outback,[37] they were to "run and note all lines by the true meridian, certifying, with every plat, what was the variation at the times of running the lines thereon noted."[38] In the late eighteenth century, this was—despite the impressive achievements of Mason and Dixon—a tall, if not impossible, order. Hutchins's surveyors' key tool was not a meridian but a circumferentor, "a simple compass fitted with sight vanes and mounted upon a ball and socket that fitted upon a 'Jacob's staff' or a tripod."[39] The Ordinance further stipulated how the townships and the lots within were to be sold. Proceeding from top to bottom and east to west, townships sold entire alternated with townships sold by lots, resulting in the characteristic checkerboard.[40] The government was happy to accept gold, silver, loan office certificates, or certificates of liquidated debts of the United States as payment.[41] Within each township, government retained the four lots numbered 8, 11, 26, and 29 for future sales. Lot 16 was reserved for the public school.[42] Ideally, each lot corresponded to a warrant or promissory note. Grid patterns, colonization,

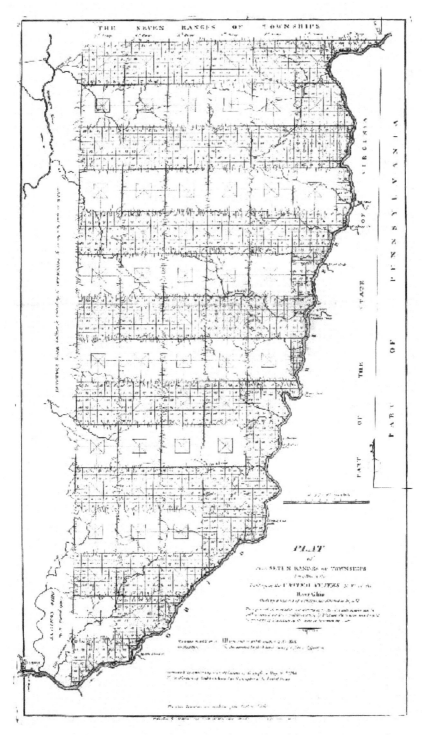

FIGURE 6-9. Mathew Carey after Thomas Hutchins, "Plat of the Seven Ranges of Townships being Part of the Territory of the United States N.W. of the River Ohio which by a late act of Congress are directed to be sold." Reprinted from *Carey's General Atlas* (Philadelphia: Mathew Carey, 1800), plate 46.

and real estate speculation coincided. "Land so marked out could be quickly and easily located by settler, banker, loan shark, and, if need be, sheriff and truant officer."[43] Once Congress entered into negotiations with speculators, the very soil of the American continent became the object of a transfer system that facilitated the circulation of real estate but did not always guarantee the optimal subdivision of the land for settlement purposes. Although the surveyors were instructed to maintain field notes on soil quality, water, and natural resources, purchasers could end up owning a swamp, a sandbank, or a piece of Native American territory. Financial and mental speculation became synonymous.[44] "The system not only made it simple to transfer land, which aided in the success of claim associations, and incidentally, that of speculators, it also contributed towards the attitude that land is a commodity."[45] While the Spanish *padròn* with its governmental doubling of grid and register reterritorialized deracinated Europeans, Congress deterritorialized the land itself. In 1785 the smallest plot eligible for sale had to be 640 acres; in order to facilitate sales the minimum was reduced first to 320, then to 160, then to 80, and ultimately to 40 acres.

The Seven Ranges survey effectively designed the entire West of the United States. The township grid is based on repeatability; it is projective, that is, it contains its own expansion up to the point of including the entire North American continent. As a cultural technique aimed at dominating space, the grid no longer appears as a potentially infinite growth of urban settlements; it is effectively cast across the land. This distinguishes the U.S. survey both from Roman centuriation and from Spanish urban planning. Settlements are no longer centers that may undergo centrifugal expansion, but cells in a homogeneous grid covering the entire territory. If Spanish colonialism was at bottom an urban affair, Jefferson's vision was essentially anti-urban, inspired by the ancient myth that cities are cesspools of vice while rural life nurtures the natural proliferation of virtues. The transformation of America into one nationwide suburb was preprogrammed. While the grids created by Roman centuriation and Spanish colonialism expanded outwards from their centers and grew toward each other in fairly haphazard fashion, the North American grid of parallels and meridians covers the entire territory. Hence the model for the latter was neither the Roman *castrum* nor the Hippodamic checkerboard but the Ptolemaic grid of latitudes and longitudes.[46] Faced with endless forests devoid of "churches, Towers, Houses, or peaked Mountains to be seen from afar,"[47] conventional European survey methods using theodolites and plane tables proved to be useless. Because the land appeared as undifferentiated as the ocean, early colonizers resorted to surveyor's chains as well as to tools used for maritime navigation, including compass, Jacob's staff, and the mesh formed by latitudes and longitudes, which defines the planisphere.[48] Transferred from the ocean to dry land, the grid encompasses the entire territory rather than merely urban space; 69 percent of the territory of the forty-eight continental states is contiguously covered by the rectangular survey.[49] Once the idea of a grid-shaped division of the land into rectangular townships and lots had been approved, bureaucratic mechanisms were put in place to ensure that the straight lines were continued west of the Ohio. "Almost nothing stood in the way. The straight lines were spread over the prairies, the foothills, the mountains, over the swamps and deserts, and even over some of the shallow lakes."[50] Never before, Catherine Maumi notes, did humans have the opportunity to confront in such brutal and violent fashion that "entity known as space."[51]

After 1796 the rectangular survey was extended to cover the remainder of the old Northwest Territory, the Southwest Territory, and other areas acquired by the United States. The township became the base unit for several governmental purposes: taxation, census, electoral districts, and road construction. It is therefore characteristic that U.S. cartography is as much based on contiguous survey plans as it is on maps. The plans created an unsparing structure that prefigured the future appropriation of the wilderness. Nothing was left untouched: The rectangular system guaranteed that no shred of land remained masterless, as frequently had been the case in the Southern territories claimed by Virginia. Both plan and projection, the uniform system of rectangular townships and sections assigned to everything—wilderness, plains, forest, or swamp—its own place. Nothing was allowed to fall off the grid.

THE THREE-DIMENSIONAL GRID

If we add a third orthogonal axis to the geodetic grid, it unfolds into a three-dimensional structure. Architecture, then, can be understood as an additional dimension of imaging, topographic, cartographic, and governmental grids. At least this is exactly how the Bauhaus architect and Gropius disciple Ernst Neufert, author of the hugely influential *Bauordnungslehre* (published in English as *Architects' Data*), saw architecture. First published in 1943 with a preface by Albert Speer, Nazi Minister of Armaments and War Production and Hitler's chief architect, it outlined a method for the complete standardization and totalization of the grid on all scales: The lattice not only connects all the buildings at a given site and determines their position and proportion, just like the grid of latitudes and longitudes, it also covers the entire globe. Neufert's grid turns navigation in smooth space into the ubiquitous paradigm of being-in-one's-place. "As on the ocean the squared grid (allows) us . . . to immediately and unequivocally determine the location of the buildings and any other installation. When built according to norm, the buildings will inevitably fit into this grid."[52] Thus Neufert's scalable planning and localizing grid anticipates the link-up of matrix screen and global coordinate system realized by Google Earth. It potentially enables on a global scale the exact location of individual buildings. On the scale of the latter, in turn, it serves to define the size and position of each object within the house, from walls, doors, and windows to stairs and furniture.[53]

If Neufert's norms were to determine our entire industrial production, then any new building would fit as seamlessly into any new settlement as any door into any door frame or any piano into any drawing room. Indeed, Neufert's own ideas fit with equal ease into the totalizing, frequently fantasmatic standardization projects that emerged in the early twentieth century. Think, for instance, of the "world format" dreamed up by Wilhelm Ostwald and Karl Bührer,[54] in which a standardized index card in a standardized filing box in a standardized cabinet in a standardized office building amounts to nothing more than the nth subdivision of a global standardization system (Figures 6-10 and 6-11).

Unfolded into three dimensions and repeated in vertical and horizontal directions, the grid does more than define the space of architecture—it turns into architecture. At the outset of the twentieth century, new materials and

FIGURE 6-10. "World Format Scholar's Library." Reprinted from Karl Bührer, *Raumnot und Weltformat*, illustrated by Emil Pirchan (Munich: Die Brücke, 1912), 24.

FIGURE 6-11. "World Format Large Library." From Bührer, *Raumnot und Weltformat*, 30.

technologies including concrete and steel frames made it possible to construct a building from the inside out. For Mies van der Rohe only a skyscraper under construction was a real skyscraper, for only so long as its sides had not been closed and covered was the steel skeleton able to make the constructive idea transparent.[55] Glass facades were therefore for Mies no more than a compromise with the inevitable. One possible construction method is to start with the smallest spatial element—the cell. Another is to focus on the steel structure and the free layout. (Le Corbusier programmatically referred to this model as a *plan libre*.) In both cases the facade is no longer part of the load-bearing structure and is thus free for almost any kind of design. "To build a uniform world from the smallest spatial cell," Walter Prigge notes, "is the rational architectural utopia of the mid-1920s."[56]

Le Corbusier, the hero of modern architecture, was one of the pioneers of an architectural dispositive in which cells (*cellules*) function as the smallest and most common element of construction. Historically, this dispositive is rooted both in the disciplinary society and in biology. On the one hand, it arises from the extensive tradition of disciplinary architecture that includes both the monastic and the prison cell. The emphasis on the cell as the smallest possible human living space reveals that modern architecture's obsession with spatial standardization is a generalization of the basic module of the disciplinary society. On the other hand, Le Corbusier is drawing on the discourse of cellular biology that identifies the cell as the basic building block of life. However, Le Corbusier's real model for cellular construction was neither plant nor prison but the machine. It is no coincidence that he developed his ideas about human-scale cells and the cell-based "dwelling machine" on board an ocean liner. Its cabins struck Le Corbusier as the optimal realization of the cellular principle under the spatial confinements of sea travel. The dimensions of the human-scale cellular dwelling, 15.75 square meters, correspond exactly to the size of the luxury quarters on his 1929 voyage from Bordeaux to Buenos Aires.

The cell had already in the early 1920s been a central component of Le Corbusier's architectural theory. Now, in 1929, as part of a series of lectures in Argentina and Brazil, he expanded the cellular concept: It is no longer a matter of designing single-family units but of planning entire cities with three million inhabitants. Here, on the continent of colonial, potentially infinite grid-shaped settlement topography, Le Corbusier proclaimed an architectural vision based on the replication of standardized, industrially prefabricated, and easily transportable dwelling cells of modular design:

> A unit of human scale: 15 square meters. . . . [T]he dwelling, the office, the workshop, the factory . . . will use new forms of standardization, of industrialization, of efficiency. . . . We shall get to the house assembled from standard components, prepared in factories, made perfect by industrialization, like an automobile body, and put up on site by *assembly workers*. . . . These methods of industrialization by standardization lead us naturally to the coming skyscrapers: its form is determined by the superposition of cells at human scale. . . . Let us multiply the standard elements of a dwelling. . . . Dwellings should not be made in meters, *but in kilometers*.[57]

Le Corbusier had the writer and aviator Antoine de Saint-Exupéry fly him across the wide plains of Argentina, Brazil, and Uruguay. The sight of the checkerboard topography of colonial cities from an altitude of 1,200 meters convinced him that "[t]his American country is dimensioned for the plane."[58] Back on solid ground, Le Corbusier drafted plans for a new Montevideo, Uruguay and a new São Paulo, Brazil (Figure 6-12). The airplane has become a design tool: From its aerial perspective the South American city of the future appears to be part of the cartographic grid. Suggesting the global grid, the future São Paulo will consist of two giant extended steel skeleton buildings that cross each other at right angles. Highways lead across the buildings; additional skyscrapers surround the intersection. Superimposed on the gridlike topography of Spanish colonial settlements, the grid of latitudes and longitudes reveals its projective nature.

FIGURE 6-12. Le Corbusier, projects for Montevideo and São Paulo, 1929. Ink on paper. Museum of Modern Art. Emilio Ambasz Fund, © 2014 Artists Rights Society (ARS), New York / ADAGP, Paris / FLC.

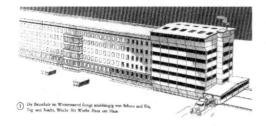

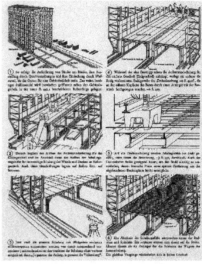

FIGURE 6-13. Ernst Neufert's railbound slipform house construction machine. Reprinted from Neufert, *Bauordnungslehre* (1943).

Rationalization by standardization is not only apparent in temporary constructions such as mass-produced barracks and containers;[59] it is also present in the visions of monumental residential constructions that can be found in Neufert's *Architects' Data*. As on an assembly line, a railbound slipform construction machine installs one cell block after the other. The three-dimensional grid continues the limitlessness of the two-dimensional topographical grid (Figure 6-13).

From this interconnectivity of grids, Le Corbusier and Neufert derived the vision of a future which in many respects is our present. The fusion of matrix grid and GPS has ensured the global presence of the operationalized deixis first conceived of in connection with the grid- and register-shaped settlements of South America. Indeed, what better way to describe some of the basic aspects of our media culture than to point to the mutual translatability of cartographic grid, topographic grid, planning grid, and imaging grid? Linked with the convertibility of these diverse grids and with corresponding scaling techniques, grids—a formidable cultural technique—have become the basis of a mediatization of space from which hardly anything can escape.

7

WHITE SPOTS AND HEARTS OF DARKNESS

Drafting, Projecting, and Designing as Cultural Techniques

DESIGNING AS CULTURAL TECHNIQUE

When architects speak of *design*, they tend to use the word in the meaning it acquired during the Florentine Renaissance: Design is *disegno*. As Wolfgang Kemp has shown, between the 1540s and 1570s the term moved from referring to a drawing or sketch produced by a schooled hand to an act of pure imagination. Following Benvenuto Cellini there was a division into a primary internal and a secondary external *disegno*, with the latter relegated to a supplementary position. Mindful of this discursive origin, it is necessary to approach the notion of design from two sides, that of the *forma* and that of the *concetto*, or idea. Design is both *lineamento* and *invenzione*, the practical realization of the *invenzione* and the inner *speculazione di menta*.[1] As a result the discourse of artistic hermeneutics came to use the verb "to design" as a synonym for "creative process."[2] When we speak of design we presume to grasp nothing less than the spiritual potential and artistic process of the creative subject. We seem to close in on the very source of creativity—that which makes the artist godlike.[3] Design in the sense of the *disegno*-discourse thus legitimizes all our delirious, self-aggrandizing fantasies centered on the artistic subject. To speak of design as a cultural technique, however, aims at something completely different. What does this change of orientation bring to light?

In order to probe the historical contingencies of the cultural technique of design, it has to be extracted from the anthropocentric origin it acquired in the Florentine discourse on art. Rather than define design as an irreducible "fundamental creative act,"[4] and thus transform it into an anthropological constant removed from history, this very definition must be analyzed as the result of discursive, technical, and institutional practices. For instance, in his introduction to the German translation of the "Introduzione di Giorgio Vasari alle tre arti del disegno cioè architettura, pittura e scoltura, e prima dell'architettura" (1568), Matteo Burioni claims that the introduction of print was an indispensable prerequisite for inventing the idea of artistic authorship, because the new means of typographic reproduction ushered in the separation between a picture's invention (*invenit*) and its practical realization (*fecit*). It is necessary, however, to go beyond such discourse-analytical clean-up operations and describe design as a recursive chain of operations.

Expressions that do little more than reinscribe the ideology of the artist's imaginary agency, by emphasizing the "active creative principle," the "real freedom of art," or the "true autonomy of drawing,"[5] must yield to an analysis of design as a cultural technique that instead focuses on material cultures, practices, and workshop conditions. Different questions have to be asked: Which mnemotechnics and storage technologies were used, what archival strategies were employed, what supporting surfaces (paper?), drawing utensils (ink, charcoal, or chalk?), and correction procedures (write over, wash off, or erase?) came into play? For the complete picture it would in addition be necessary to investigate to what degree the rhetoric of artistic freedom and autonomy is subject to discursive rules that arise from the institutionalization of *disegno* as a discipline and, on a deeper level, from the emergence of the governmental techniques of the early modern state[6]—though these aspects of the discussion are beyond the scope of this chapter.

Speaking with Hegel, the freedom or openness of the manual drawing is not the most immediate but the most mediated. It does not stand at the beginning of the creation of paintings, buildings, or machines, but at the end of a protracted disciplinary process that imposes discursive rules on and codes the relationship between hand and eye. To invoke the basic, irreducible act of artistic creation is to ascribe agency exclusively to human actors, which is as ignorant as celebrating the "scientific mind" as the prime source of the early modern scientific revolution. These misguided notions display obvious similarities; hence the rethinking Bruno Latour demanded from the history of science applies equally to the history of art and architecture. First, what was attributed to the "scientific mind" (or the artistic imagination, respectively) has to be ascribed to "the hand, the eye and the signs." Second, signs are not to be treated as signs but as media.[7] This opens up a new path into the history of design, a history that eschews the empowerment and celebration of the creative ego in favor of the exteriority of thinking, forming, and shaping.

According to Latour, the ability to design the incomplete, let alone the impossible, on paper is indebted to "immutable mobiles." Once we view design itself as an "immutable mobile," the semiotics of self-presence ("here the artist is fully present")[8] are replaced by qualities such as mobilizability, combinability, scalability, superimposability, geometrics, and so on. Why was central perspective such an important invention? Because objects could be turned and moved while staying the same. Perspective generates optical consistency. The prints of Georgius Agricola, for instance, depict objects or fragments of an explosion on one and the same sheet of paper in different scales and from different angles and perspectives. The "optical consistency" allows for mixing the parts.[9]

With this in mind we also need to go beyond the evolutionary framework of Wolfgang Kemp's conceptual history of *disegno*. What is most revealing in the decades analyzed by Kemp (1540–1570) are the many contradictory statements, for instance, by Vasari. They do not reveal a teleological development of *disegno* leading towards a mental projection, but rather an increasingly problematic academic knowledge of the fact that a drawing is not merely the external supplement of an internal idea. For Vasari *disegno* is, first, something akin to synthetic reasoning; second, the ability to recognize scale in natural and artificial bodies (plants, buildings, sculptures, paintings); and third, the "visible expression and declaration of our inner conception."[10] With regard to the latter, *disegno* requires "that the hand, through the study and practice of many years, may be free and apt to draw and to express correctly, with the pen, the silver-point, the charcoal, the chalk, or other instrument, whatever nature has created."[11] Vasari's wavering between clearly separating *invenzione* and *designo* on the one hand and identifying the two on the other indicates that he was still attached to a notion of an inventive potency or operational efficacy that rests with the codes and media of the drawing itself. One hundred years later, Leibniz put a great effort into designing an *ars inveniendi* based on self-acting signs (characters).[12]

To understand design as a cultural technique, then, involves our subjecting it to a historical apriori of technologies, materialities, codes, and visualization strategies rather than attributing it to some ineffable act of creation. The prototype of the architect is not the demiurge, whose superhuman act of strength and volition creates the world from primeval chaos by separating the aesthetic Apollonian line from the libidinal Dionysian night, but the artist-engineer. If we are to present the double nature of *disegno* as *lineamento* and mental projection (*speculazione di menta, invenzione*) as a history of cultural techniques, hollow phrases invoking "artistic creativity" must yield to an analysis of concrete sign practices.

LEONARDO DA VINCI: DESIGNING AS AN EXPERIMENTAL SYSTEM

First and foremost we need to call into question the distinction between technical and artistic design. Leonardo da Vinci's water studies may serve as an example: Not only do they reveal how untenable this distinction is; more importantly, they show that the *invenzione* of the design process is rooted in an experimental system consisting of a diverse arrangement of things, codes, and media of inscription.

While Alberti saw water as a topos of *imitatio* created by nature itself,[13] Leonardo appreciated it as an ever-changing body, a dynamic element that constantly generates forms, especially in the shape of swirls and eddies. Using quill and ink to sketch such turbulences in smooth space, Leonardo is in effect translating what no line can capture into a code of more or less prominent lines. There is no line without a nonline. The line can provide water with a distinct shape, and where there is no interim space it can let this closed body unravel. Drawing reveals the inventive potential of flowing water. Leonardo's studies of water in motion transform into preliminary studies for the depiction of hair (Figures 7-1 and 7-2). In his notebooks this design procedure is articulated as an instruction: "Watch the movement of the surface of the water, how like it is to that of hair, which has two movements, one following the undulations of the surface, the other the lines of the curves; thus water forms swirling eddies."[14] The drawing becomes a medium between water and hair. Hair, the three-dimensional real-life object that comes closest to lines on paper, naturalizes the rendition of smooth water. But where in this chain of operation is the moment of invention? Which participating thing, what organ of perception or execution is responsible? The eye of the artist? His hand? Or maybe the water, the quill, the ink, or the paper? Answer: none of the above, though they all participate in the design. Rather, the invention emerges from transfer operations, the transfer of the form observed in the water onto the medium of drawing and the unconcealing of a specific substantiality, a specific materiality emerging from this *lineamento*.

None of this is happenstance: The act of designing is itself guided by design. Before the artist-engineer grabs his quill, his hands take hold of heavier items. The Codex Leicester sketchbooks reveal Leonardo's efforts to elaborate an artistic vocabulary of fluid motion articulated by basic geometric bodies: flat cubes, cylinders, cuboids, cones (Figure 7-3). In the beginning, then, is not the unique ingenious idea but the series. The series clarifies that we are dealing with an experimental system. Water does not articulate itself. Leonardo insists that water, when left to its own design, always heads for a state of rest that is realized in the shape of the ocean. The ocean is the maximum entropy of all articulations, the ground that has absorbed all figures.

However, the basic geometric bodies used by Leonardo to represent elementary swirls and eddies on paper are media rather than elements of articulation. As becomes evident in Leonardo's unflagging interest in the cultural techniques of hydrology, the ornamental figuration of water always articulates itself by means of something else: piers and abutments in rivers, canals, openings through which water pours into basins, not to mention his numerous sketches of water gushing along and around wall fragments and stone plates. Leonardo, in short, is interested in boundaries and interfaces. As his antediluvian landscapes indicate, he did not see surfaces as polygons of Euclidean planes but as boundary planes shaped by the interaction between moving elements. In other words, these boundaries are the traces of forming or deforming geological and climatic processes. A plane surface is for Leonard the trace of a leveling, a convex surface the trace of a filling, and a concave surface the trace of a hollowing by water currents and shapes. Drawing technique and cultural technique tend to coincide, for example, when Leonardo sets out to study how water eddies start to model an initially concave surface (Figure 7-4). Underneath the

FIGURE 7-1. Leonardo da Vinci, four studies of swirling water, reminiscent of loosely braided hair; c. 1513. Pen-and-ink drawing on white paper, 152 × 213 mm. Royal Collection Trust / © Her Majesty Queen Elizabeth II, 2012.

FIGURE 7-2. Leonardo, sketches of a female head for *Leda*, a lost painting; date disputed. Pen and ink over black chalk on white paper, 200 × 162 mm. Royal Collection Trust / © Her Majesty Queen Elizabeth II, 2012.

sketch he notes (from left to right): "fills the ground / levels the ground / hollows out the ground."[15]

When Leonardo draws water in motion, he is not only drawing its ornamental shapes but also observing design as something that takes place within the experimental system (rather than out there in nature). By virtue of its fluidity and the innumerable shapes it assumes, water more than anything else

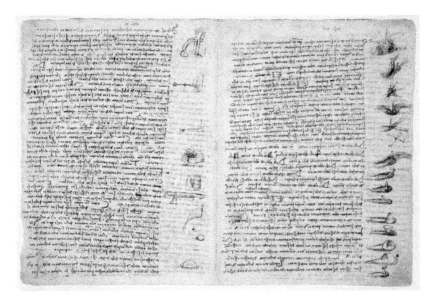

FIGURE 7-3. Leonardo, study of the various turbulences produced by differently shaped objects in flowing water. Codex Leicester, sheet 14A, folio 14r. Private collection. © Seth Joel / Corbis.

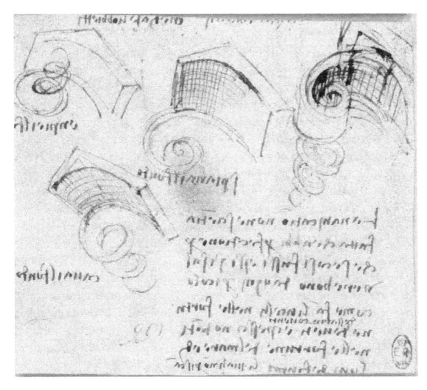

FIGURE 7-4. Leonardo, three studies of water swirls emanating from a concave surface; undated. Ink on royal white paper, 88 × 101 mm. The legends read, from left to right: "fills the ground / levels the ground / hollows out the ground." W. 12666r. Royal Collection Trust / © Her Majesty Queen Elizabeth II, 2012.

FIGURE 7-5. Leonardo, sketch of water flowing from a rectangular opening into a bowl; 1507 or 1509. Pen and ink, 290×202 mm (lower half). W. 12660v. Royal Collection Trust / © Her Majesty Queen Elizabeth II, 2012.

resembles imagination—it is a designing, inventive entity. By subjecting water to certain arrangements, the engineer can perpetuate (though not arrest) its swirling motions, while the artist, in turn, is able to wrench an image from this liquid spirit. We do not start with a ground against which the figurations stand off; rather, ornamental and grotesque figurations create the ground from which they, in turn, emerge. One revealing example is Leonardo's sketches of water falling from an opening into a basin already filled with water (Figure 7-5). At the point of intersection of the two bodies of water, one at rest and one in motion, ornamental articulations arise—the typical ornate bubbles and frizzles—from the midst of which plantlike shapes emerge. In the Codex Leicester, Leonardo describes the foam resulting from water plunging into a water-filled basin: "The air which is submerged together with the water which has struck upon the other water returns to the air, penetrating the water in sinuous movement, changing its substance into a great number of forms."[16] When you compare this sketch with the *Diluvi* series, one recognizes the very same process but in reverse: The sketch was concerned with the articulation of figures resulting from the interaction between receiving and inscribing bodies of water or ink, whereas in the *Diluvi* it is a matter of disarticulating figures, of dissolving them into the swirls and turbulences generated at the intersection of deluge and dry land.[17]

SYMBOLIC WORLD ORDERS

The Kabyle house, as described by Pierre Bourdieu, represents a perfect example of a culture based on distinctions.[18] It is a structuralist's dream: The space of the house is an arrangement of homologous oppositions—fire/water, cooked/raw, above/below, light/shade, day/night, male/female, fertilizing/prepared for fertilization, culture/nature (Figure 7-6).[19] The structural differentiation and semanticization within the house repeats the difference between inside and outside. The opposition between cosmos and house is derived from the fact that one side of this opposition, the house, is itself articulated according to the same principle (a: b / b_1: b_2). The sexual opposition, which is at the same time a cosmic difference, is reflected by the architecture of the house, most prominently in the main pillar and the master beam supporting the roof. There is no space for a design process; the Kabyle house cannot be designed because it has always already been designed. It reflects the cosmic order of nature and culture, man and woman, inside and outside, and so forth, and it reflects this reflection.

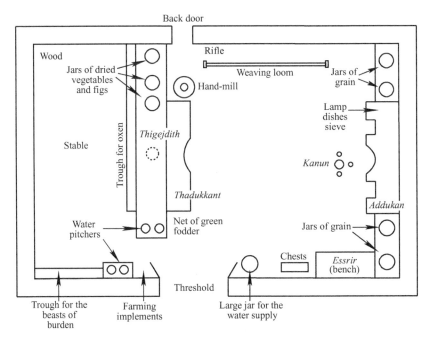

FIGURE 7-6. Kabyle house plan. Reprinted from Bourdieu, "The Kabyle House or the World Reversed," in *Algeria 1960*, p. 134.

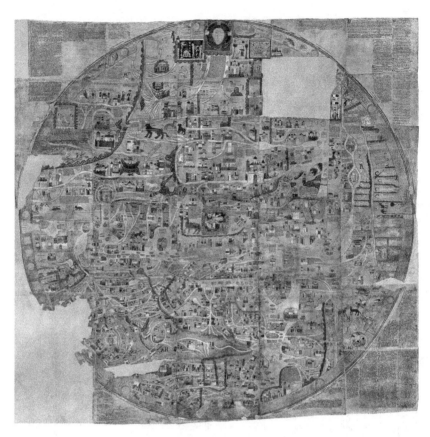

FIGURE 7-7. Ebstorf map, thirteenth century. Reprinted from Ute Schneider, *Die Macht der Karten*, 24–25.

Much like the Kabyle house, medieval *mappae mundi* represent symbolic world orders that are in essence nothing but spatial encodings of narratives (Figure 7-7).[20] The medieval *mappae mundi* are more or less based on T-O maps, that is, pictorial charts in which space is a mesh of topoi. Their topography is a form of "place-writing" that does not communicate any geographic knowledge but spatializes narratives of salvation. T-O maps enable multiple subject positions for salvational, mythical, and biblical stories, thus allowing many tales and voyages in and between texts to emerge. An endless number of stories may be generated that nonetheless remain subject to the grand narrative of salvation. In such a world, too, the notion of design makes little sense. The world of the *mappae mundi* has already been designed, and man has already been thrown into an already interpreted, thoroughly symbolized and encoded world. To return once more to the Kabyle house: Here, too, space and its divisions are revealed by many narratives. Apparently the house does not exist without tales, riddles, and proverbs that provide the foundation for the functions of the edifice and its rooms (who may do what in which place at what time). Spatial codes cannot be separated from structural symbols; the latter, in turn, exist solely in the medium of everyday language. Here no white spots would allow a subject to come into being.

CARTOGRAPHIC SELF-DESIGN

The modern European subject is a project designed by designs. Its design is of no less than planetary dimensions. Painted portraits have time and again captured this gesture of self-projection: The confident European subject faces the observer and points with his finger at a map of—say—New Zealand.[21] This still occurs at the end of the nineteenth century, just as the last white spots are about to vanish from the maps. One of those subjects around 1900 demonstrating the cartographic dimensions of self-design goes by the name of Joseph Conrad. At the beginning of *Heart of Darkness*, Marlowe, the frame narrator, recounts a childhood memory:

> Now when I was a little chap I had a passion for maps. I would look for hours at South America, or Africa, or Australia, and lose myself in all the glories of exploration. At that time there were many blank spaces on the earth, and when I saw one that looked particularly inviting (but they all look that) I would put my finger on it and say: "When I grow up I will go there."[22]

In his "Personal Record" of 1908 and then again in 1923, Conrad rewrote Marlowe's recollection as his own childhood memory: "One day, putting my finger on a spot in the very middle of the then white heart of Africa, I declared that some day I would go there."[23] The encoding of the unknown in cartographic space allows the Pole Josef Korzeniowski to design a literary "I" that incorporates this empty space into his own life as a place of projection. A few years later, this cartographically facilitated projection is inscribed into his own life projection. Marlowe's childhood, in other words, is no autobiographical reminiscence of Conrad's childhood; Marlowes's childhood is Conrad's retroactively written "I." First there is the map as the design for the literary "I," then the literary "I" as a design for one's own ego.[24] These are the two exemplary operations of the projective recoding of the occidental subject as a design or "projectile." The decisive moment of the drawing that turns into a *disegno* is made up of techniques of projection and project-making.

DESIGN AS PROJECT(ION)

To describe design as a cultural technique means to distance oneself from the Florentine reading as *disegno* and instead conceive of it as project, projection, or projecting. In "The Age of the World Picture" (1938), Martin Heidegger defined "projection" (*Entwurf*) as the basic procedure of modern scientific research.[25] More precisely, research—as "[t]he essence of what we today call science" (118)—is defined as a "procedure" (*Vorgehen*; 118.). This "procedure" is not merely to be understood as a method, but also, and quite literally, as moving-forward (*Vorwärtsgehen*), a setting-out into the unknown, a voyage of discovery, conquest, and research eager to seize and apprehend the unknown in the shape of a picture. "The fundamental event of the modern age is the conquest of the world as picture" (134). This procedure, however, this designing of oneself with a view toward something still unknown, requires a preparatory draft (*Vorzeichnung*), a projection defined by Heidegger as "a certain ground plan of natural events" (118; translation emended).[26] The question arises of how we can transfer Heidegger's notion of projection to the arts of *disegno*, that is, to architecture, painting, and sculpture. When Heidegger speaks of projection (*Entwurf*), he is not referring to the specific draft or projection of a given composition; rather, he has in mind the basic projection underlying the design of any such composition. Heidegger's understanding of projection and his specification of it as "ground plan" is reminiscent of Leon Battista Alberti's central perspectival grid; that is, it recalls the division of the floor or the canvas itself into orthogonals and transversals according to the rules of perspectival foreshortening. It is important to remember that for Alberti, the importance of the basic grid resides less in the correct progression of diminishing distances between the transversals than in its fundamental contribution to the overall composition. In *On Painting* he wrote, "This whole procedure of subdividing the pavement pertains in particular to that part of painting that, in its place, we shall call composition."[27] *Compositio* is a term from rhetorics that Alberti was the first to apply to painting.[28] Indeed, Alberti dissects a painting in much the same way as a rhetorician analyzes a sentence. *Circonscriptio*, a term that in Alberti's treatise refers to outlining the contours of a body, was used by rhetoricians to indicate a period. Together with the *luminum receptio*, the correct distribution of light and shade, the *circonscriptio* makes up the *compositio*.[29] Because they were unable to subdivide the grid correctly, there is no "*historia* by the men of the past" that is "composed to perfection."[30] While the recent English translation of *On Painting* by Rocco Sinisgalli retains the original word, the German translation prefers to render *historia* as "procedure" (*Vorgang*), thereby unwittingly establishing a connection to Heidegger. Exploratory research and artistic invention converge, literally and conceptually, in the notion of a procedure that is projected by a basic plan.

This, in turn, suggests a systematic as well as historical connection between the perspectival grid and the Ptolemaic grid of latitudes and longitudes used in Mercator projections. Indeed, in 1975 Samuel Edgerton proposed that the principles of Alberti's linear perspective were already contained in the third mapping method discussed in Ptolemy's *Geographia*.[31] In the early 1400s, after the Byzantine scholar Manuel Chrysoloras had brought a copy of the *Geographia* to Florence, the city turned into a center for cartographic and geographical studies. Dom Pedro, brother of Henry the Navigator, visited Florence in 1428, presumably to buy maps. Around 1412, Cardinal Pierre d'Ailly wrote two commentaries on Ptolemy's *Geographia* that eventually found their way into the library of Christopher Columbus.[32]

Ptolemy's third method of projection represents the *oikumene*, the inhabited world, as seen from an individual human eyepoint (Figure 7-8). Earth appears as a circle, with a vertical axis running through the center point connecting the poles Π-P and a horizontal axis representing the equator. Point Σ located above the equator indicates the point at which the vertical axis is intersected by the latitude running through Syene (the latitude of Syene—today's Assuan in Egypt—marks the middle between the northern and southern boundaries of the *oikumene*). This point is directly opposite the distance point or point of

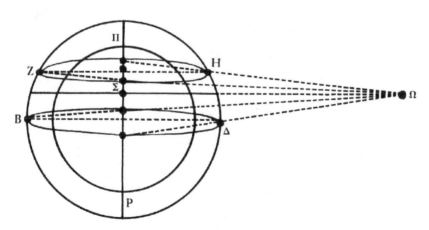

FIGURE 7-8. Ptolemy's third projection method. Reprinted from Edgerton, *The Renaissance Rediscovery of Linear Perspective*, 109.

sight (Ω), which thus marks both the center of the observer's field of vision and the center of the *oikumene*. The meridian connecting North and South Poles as well as the Syene latitude (Alberti's *punctus centricus*) appear as straight lines, whereas the remaining latitudes appear as concave lines which, if extended around the globe, form ellipses. The latitudinal rings above Syene thus appear as perspectivally foreshortened rings seen from below, while those beneath Syene appear as seen from above.[33]

We arrive at a picture of the globe as if reflected in a convex mirror. The same method of projection was used by Parmigianino for his *Self-Portrait in a Convex Mirror* (Figure 7-9). Maybe the rounded shape of the picture, which was painted in accordance with Ptolemy's third method of projection (showing the latter to be a "non-Euclidean" special case of Alberti's linear perspective), does not simply refer to a shaving mirror,[34] but to the globe. In that case, Parmigianino's self-portrait reveals the planetary self-projection of the modern subject.

FIGURE 7-9. Parmigianino, *Self-Portrait in a Convex Mirror*, 1524. Oil on wood. © Kunst historisches Museum, Vienna.

Projection involves moving from body to image (Ptolemaic maps were called *pitture* in Florence). This transition occurs in the cartographic projections, architecture, and artisanal workshops of the Italian Renaissance. In different though maybe connected ways it also occurs in the fashioning of the central perspective. What unites all these practices and procedures is the basic act of conquering the world, now turned into a picture, by means of projective graphic operations that turn the subject into a stage of representation. For Leonardo the eye is the active principle of design that connects the artist's workshop with the navigators' globe:

> The eye is the master of astronomy. It makes cosmography. It advises and corrects all human arts. . . . The eye carries men to different parts of the world. It is the prince of mathematics. . . . It has created architecture, and perspective, and divine painting. . . . It has discovered navigation.[35]

The eye is both origin of the visual rays and the vessel that carries explorers to the farthest reaches of the earth. "Today the painting—and tomorrow the world."[36] Alberti's device—his *imprese*—was a winged eye.

It is the intersection of central perspective and navigation that unleashes the specifically modern European dynamics of invention. Transatlantic discoveries reveal with particular clarity what the word *invenzione/invenire* has come to mean in the sixteenth century. On the one hand, *invenire* refers to the act of finding something, as it already did in antiquity, as well as to the notion of invention it acquired in the *disegno* discourse. On the other hand, it refers to discovery with an audible undertone of downright seizure. In Hugo Grotius's famous treatise *The Free Sea*, discovery involves more than merely visual apprehension, "for to find is not to see a thing with the eyes but to lay hold of it with the hands"—in increasingly violent fashion, we must add. *Invenire* and *occupare* are synonymous.[37] The two-dimensional medium triumphs over its three-dimensional counterpart; the image vanquishes the body. Imperialism is applied planimetry. "Sovereignty belongs to the one who decides on flattening."[38]

Media capable of addressing the unknown and unfinished generate and frame "disinhibited subjects" whose life-goals have been moved from a heavenly up-and-beyond to a terrestrial over-there.[39] Antonfrancesco Doni's *speculazione di mente*, which he used in 1549 to characterize the *disegno*,[40] must be understood in its philosophical as well as its economic dimension. The design promotes speculative action; to design is itself a speculative act, to operate at one's own risk: It indicates the emergence of subjectivity from operations of borrowing, investing, planning, inventing, betting, reinsuring, and risk-spreading. One gambles on the return of capital invested in the future or in a transatlantic beyond. The white spots, the unknown territories noted on maps since around 1500, can be converted into speculative profits. Maps become the study for risky endeavors.

LOXODROMES

As Edgerton emphasizes, Ptolemy's third method of projection was never put to practical use. There is a simple reason for this. Since the distances between the Mediterranean coasts are not that extensive, the discrepancy between the compass heading and the ship's actual course is negligible, but once navigators ventured out into the Atlantic and Indian Oceans they had to take into account the earth's curvature. A map with simple linear headings was insufficient; it became necessary to project the third dimension into the second.[41]

Gerard Mercator realized that the reason why seamen frequently provided him with false data was their erroneous assumption that they would sail in a straight line when following a rhumb line. He noticed that a ship that keeps heading toward one and the same point of the compass describes a curve known as a loxodrome (or spherical helix). To plot the ship's course in a straight line therefore requires a special method: the orthomorphic cylindrical projection. The distance between the latitudes increases in the same ratio as the distance between the spherical meridians diminishes. Taking this into account allows navigators to chart the ship's course, that is, the loxodrome, as a straight line.[42] The voyage becomes a matter of design; the world is encountered as a project. The ships itself turns into a projectile: It stays its course by means of operative calculations performed by the graphic surface of the grid map. The Mercator projection becomes part of the operations that keep the ship on its "straight" course. The grid becomes a "fundamental tool"[43] for the visions of the discoverers.

OPTICAL CONSISTENCY BETWEEN WORKSHOP AND GLOBE

Ptolemy not only pioneered the projection of the spherical surface onto the plane surface of a map, but also the method of dividing the latter into a grid composed of latitudes and longitudes, which served to reduce "the traditional heterogeneity of the world's surface to complete geometrical uniformity."[44] Though Ptolemy's map depicted no more than the *oikumene*, that is, the known world, the grid implied that the *oikumene* only constituted a part of the world's entire surface. There were longitudes beyond the Pillars of Hercules and the Canaries that marked the westernmost boundary of the known world and Ptolemy's map. Unlike medieval *mappae mundi*, Ptolemy's maps embody a grid system.

Very soon that which the map happens to depict will be no more than a contingent part.

Of course the grid had been known in Europe since antiquity. Starting with Roman *agrimensores*, surveyors had used it to plan settlements or—as Brunelleschi had done in Rome—measure buildings.[45] However, following its rediscovery in the early fifteenth century, the Ptolemaic grid was articulated in terms of fixing proportionalities, especially the proportionalities of distances, which also happens to be the key aspect for the construction of central perspective.

> No matter how the grid-squared surface is shrunk, enlarged, twisted, warped, curved, or peeled from a sphere and flattened out, the human observer never loses his sense of how the parts of the surface articulate. The continuity of the whole picture remains clear so long as he can relate it to at least one undistorted, modular grid square.[46]

In other words, the Ptolemaic grid reveals the cultural technique of optical consistency, and it is precisely this cultural technique that mobilizes drawings as tools of design in the artisan workshops of the Florentine Renaissance. Is it mere coincidence that at the very moment in which the *Geographia* was copied and disseminated in Florentine *scriptoria*, Masaccio stopped using the old *sinopia* technique for the underpainting of his *Trinity* fresco in the Dominican church of Santa Maria Novella?[47] In order to transfer the face of the Virgin Mary from drawing to wall, Masaccio used a grid directly applied to the freshly plastered surface (Figure 7-10).

In her detailed study of Italian Renaissance workshops, Carmen C. Bambach reconstructed the techniques that connected drawings to designs, plans, and sketches. Spending fifteen years clambering up and down the scaffolds erected to restore Renaissance frescoes, she was able to detect a large number of clues indicating mechanical transfers, especially the use of *velo*, *spolvero*, and *calco* (also called *calcare*, *ricalcare*, or *incisione indiretta*). The *velo* or *velum*, a device Alberti claimed to have invented (a claim later confirmed by Vasari), consists of a thin cloth placed between the artist and the object to be copied. According to Alberti, the *velo* teaches the artist to perceive the subtleties of relief, perspective, proportion, and outline.[48] It is thus a means of disciplining eye and hand. As Bambach was able to show in the case of Masaccio's *Trinity* and other studies and models (e.g., by Sandro Botticelli, Michelangelo, Raffael, Jacopo Tintoretto, and Paolo Ucello), the *prospectivi*—Cristoforo Landino's 1481 word for "masters of perspective"—used the squared grid from about 1420 on.[49] Edgerton goes so far as to claim that the *velo* was "the most interesting adaptation of Ptolemaic 'space structuration' to the practice of painting."[50] Leonardo and Albrecht Dürer constructed machines that used the *velo* as a medium for perspectival

FIGURE 7-10. Masaccio, *The Holy Trinity* (detail), 1425–27. Fresco. Santa Maria Novella, Florence.

depictions. Traces of the *spolvero* technique have been found on numerous frescoes. It consisted in dusting the drawings with charcoal powder which then passed through needle-pricked holes (a technique Dutch painters called *griffeln*) onto the wall, leaving fine dotted lines that helped the *maestri* to execute the painting (see Figures 7-11, 7-12, and 7-13). In the case of the *calco* the drawing is directly etched into the still-soft plaster using a sharp stylus. Traces of such incisions can be found in Masaccio's *Trinity*, among other works. The "invention,"

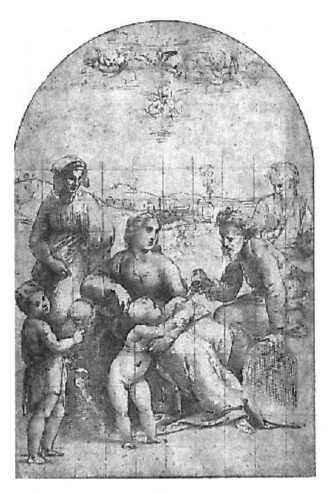

FIGURE 7-11. Raffael, design drawing for *Holy Family with the Pomegranate*, c. 1507.
Pen on top of stylus marks and chalk, gridded. Musée des Beaux-Arts, Lille.
Reprinted from Westfehling, *Zeichnen in der Renaissance*, 260.

then, can be deciphered as the trace of a mechanical transfer operation involving a multitude of assistants (*manuali*), fresco painters (*maestri pratichi a lavorare a fresco*), painting masters (*maestri pictori*) who specialized in individual elements (curtains, skies, clouds, ornaments, backgrounds, wax models), and *maestri* responsible for transferring the cartoons (*cartones*) onto the wall. Around 1540 Italian artists viewed the cartoons as the most important stage of the composition. Such artistic practices, Bambach argues, can be reconstructed from the "material culture pertaining to the workshop."[51] However, to reveal the techniques

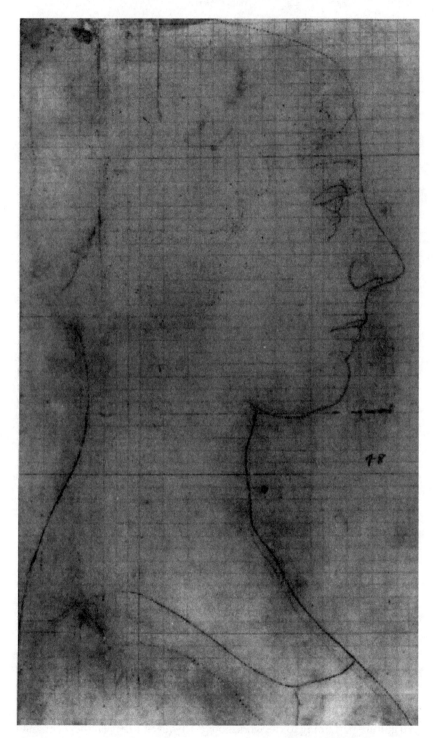

FIGURE 7-12. Workshop of Leonardo, contour cartoon for a portrait of a woman in profile. The image was drawn on a grid of pen and ink, and black coal dust was rubbed into the perforated contour lines. CBC 351, Royal Library, Windsor Castle, inv. 12808. Reprinted from Bambach, *Drawing and Painting in the Italian Renaissance Workshop*, 278.

FIGURE 7-13. Raffael, perforated cartoon for an allegory entitled *The Knight's Dream*, c. 1502. Recto and verso have been rubbed with black coal dust. CBC 231, British Museum, London, 1994-5-14-57. Reprinted from Bambach, *Drawing and Painting in the Italian Renaissance Workshop*, 14.

involved in actualizing the design processes is not to deny the genius of the Renaissance masters, "but rather to understand how fundamental a tool drawing was to their vision."[52] To (mis)quote Nietzsche's (in)famous claim: Our design tools are also working on our imaginations.[53]

We therefore need to qualify the claims Alberti makes in *On Painting*. It appears that the veil, which Alberti professes to have invented himself, was already in use at the time when he first penned his treatise on painting. Francis Ames-Lewis suspects that Alberti's advice to fellow artists to use a *velum* reproduces "an accepted technical practice in the circle of Brunelleschi and Masaccio."[54] At least it appears to have been used in Masaccio's *Trinity*, the first painting to be executed in correct perspectival fashion. In addition Alberti emphasized the usefulness of subdividing the schemes (*modelli*) "into [a network of] parallels, so that in a work [for the] public, all objects, . . . taken from quasi personal sketches, are arranged in appropriate positions."[55] Hence, for the alleged father of the *disegno* the scheme was first and foremost an "immutable mobile," a transfer and projection tool. His use of terms like *concetti* and *modulos* indicates a recoding in the course of which mechanical tools and rhetorical examples transform into manifestations of creative imagination. The fact remains, however, that drafts and sketches were only able to turn into documents of invention because they separated the completed work as the merely mechanical execution from the draft as a medium aiming at mechanical imprint or incisions. In the course of the differentiation between manual media the design is released. It is, as it were, set free to become a product in its own right. Before the line can be appraised as the expression of an idea, it prescribes the activity of the stylus, which as part of the *spolvero* technique will puncture it. Inasmuch as line and puncturing are connected in an instrumental way, the line becomes part of a culture-technical chain of operations. Because it is used for the purposes of reproducibility, it can transform into a trace of the original—an individual and creative originality that manifests itself in the unfinished, that which is still open for future alteration. The emergence of the draft from the preparatory drawing, which was used for the mechanical transfer onto the wall, can be traced with the aid of Michelangelo's and Leonardo's Florentine drawings. In 1503/04 Leonardo and Michelangelo were commissioned by Piero Soderini to decorate the Sala del Maggiore Consiglio in the Florentine Palazzo della Signoria. Both artists' preparatory drawings and cartoons for the battle frescos were subsequently studied by budding Florentine artists as models for drawings as *disgegno*.[56] Techniques of scaling, transferring, and impressing give birth to the *idea*.

To merge drawing and design, in the sense of preparatory drawing or plan, means to keep the drawing open for future alterations: the drawing contains and projects the possibility of a future completion. Yet it is not only a space of optical consistency that provides a calculable space devised in advance for its own future completion. The drawing is a medium that enables mobilizing effects; indeed, it facilitates the creation of workshops in which artists can focus on the design part and leave the execution to collaborators. The drawing is a guidance and control tool, because it allows for the correction of mistakes and prescribes corrections.[57]

EXITING THE PROJECT

It is therefore by no means self-evident that the "sketched, the crude, the linear, always [refers] to something unfinished, to something that can still take shape,"[58] as asserted by an anthropocentric art history that is as blind to media as it is enraptured by individual artists. The unfinished, that which may still become something and serve as a (grid-based) projection surface for future subjects, emerges from the mechanical transfer or projection these sketches were used for in Italian workshops. Yet how difficult it was at the beginning of the early modern age for the unfinished to emerge as the unfinished, can be glimpsed from a 1502 portolan map by Juan de la Cosa (Figure 7-14).[59] In the west the map depicts the Atlantic and parts of the newly discovered continent. The Caribbean and portions of Brazil have been more or less charted. The unknown, located between the westernmost boundary of the known world and the very edge of the map, is filled with green and brown. This incomplete draft is a projection containing both an exit from the map and an entry into images. Where the westernmost boundary of the known world intersects the thick red equatorial line, de la Cosa added a cartouche depicting Saint Christopher with the infant Jesus on his shoulders stepping into a river whose other bank is beyond the horizon. Christopher's entry into the *okeanos*—a river with no other side— is the exit from that which is contained within the draft at precisely the point at which a map should present itself as unfinished. Here, at this boundary, the portolan map offers a way out of itself and a re-entry into the old *mappae mundi*. It transforms from a world-as-design back into a symbolic world order, rather than into the unfinished open of linear projection.

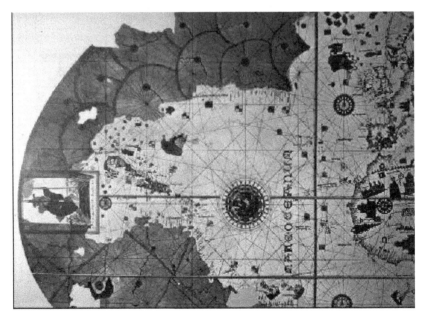

FIGURE 7-14. Juan de la Cosa, world map, 1502. Museo Naval, Madrid. Reprinted from Wigal, *Historic Maritime Maps, 1290–1699*, 54.

The open and unfinished emerges with techniques that enable us to write the merely possible and operationalize the nonrealized. In other words, we are dealing with media whose codes contain variables. Spain and Portugal, the great maritime empires of the sixteenth century, entered the age of cartographic processing in which all the maps used on their ships were based on a constantly revised and by definition interminable master map, the *padrón real*. With the *padrón real* the ongoing improvement of the design of the world became the basis for all actual representations of the world.

> To bring order to all things . . . we hereby command that a royal map [*padrón real*] be drawn, and to ensure that it be drawn with the utmost accuracy, we hereby command our officials in the Casa de la Contratación in Seville to assemble the most skilled navigators . . . and that in the presence of you, the named Amerigo Vespucci, our *Piloto Mayor*, a map of all the lands and islands of the Indies that so far have been discovered be drawn . . . and that, pending the assent of the *Piloto Mayor*, a general map be drawn which shall be named the Royal Map [*padrón real*] whose purpose shall be to govern and guide [*egir e gobernar*] all navigators . . . and no navigator shall make use of a map which is not a copy thereof [*que fuese sacado por el*].[60]

"America is a cartographic revolution, the cartographic revolution is America."[61] In other words, America is the ontological result of a planetary design. The *padrón real* was a medium of knowledge unknown to the Middle Ages. It unleashes design practices triggered by the Florentine short-circuit of Ptolemaic grid and *velum*. The *padrón real* is a virtual map that itself is never used as a map because it is in a permanent state of design. The medial practices that transform it into the secret standard for all navigational charts of the Spanish-American empire defines it as incomplete.[62] It charts the ever-changing level of permanently provisional knowledge in need of ongoing revisions. The *padrón real* stores virtual data; it is a control medium that not only steers individual cybernetic machines (ships) that manage to stay on course by means of a loop comprised of location determination, map matching, adjusting rudder and rigging known as navigation, but also guides its own improvement. For wherever the voyages inspected by the *piloto mayor* may lead, their return journey has only one address, the Casa de la Contratación in Seville, seat of the *piloto mayor*. Seville is the eye Leonardo spoke of, the one that sees the whole world.

> Furthermore, we hereby command that upon their return to Castile all navigators in our employ . . . who have come across new lands, islands, bays or

harbors, or anything else worthy of being added to the aforementioned *pa-drón real*, submit a report to you, the *Piloto Mayor*, and your officials [*dar su relacion a vos*] so that all can be entered in its right place in the *padrón real*.[63]

The *padrón real* is the "mutable immobile" underlying all "immutable mobiles." Its global projection is based on the homogeneous Ptolemaic grid of longitudes and latitudes covering the globe and equipping all that is still unknown with a priori known addresses. White spots are the result of a cultural technique which renders addressable everything that is rather than is not. Henceforth data are the destination (*Geschick*) of their addresses.

CONCLUDING REMARKS ON THE GENESIS OF DESIGN

Even though Edgerton later revised his 1975 thesis on the origin of central perspective from Ptolemy's third cartographic method, the historical fact remains that the early Florentine Quattrocento witnessed a momentous short-circuit of paper and global surface that gave a radically new meaning to the term *invenzione*. The Ptolemaic grid, which ontologically formats the still unknown, undiscovered, and unfinished as a destiny of addresses and transforms them into objects of speculation, is a grid projection deployed on a planetary level. For Alberti, it was the decisive component of the composition process (*historia*), one which could be replaced by the veil and other transfer and scaling grids employed in Italian workshops since the fourteenth century (for instance, in the workshops of Ghirlandaio or Raffael).

According to Edgerton the Ptolemaic grid is "a skeletal geometric key to the link between Quattrocento cartography and the paintings which gave birth to linear perspective."[64] What does this mean for our account of design as a cultural technique? The merger of artistic *compositio* and draft was based on the superimposition of the Ptolemaic grid and the grid techniques of the fresco painters, of whom we know that they were familiar with the invention of central perspective by Brunelleschi and its theoretical elaboration by Alberti. Design, then, did not originate in an ingenious *uomo universale* who in a mighty feat of will managed to design design,[65] but in the convergence of two cultural techniques that flourished in fourteenth-century Florence: the cartographic grid of latitude and longitudes and the grid-based techniques of scaling, proportion, and transfer employed by Renaissance artists. Historically, the meaning that *disegno* acquired much later on can be traced back to the amalgamation of the differing concepts of the "open," which in conjunction with the two cultural techniques found their way into drawings. Emerging from a field of tension between artisanal and cartographic projection techniques, the artistic design merges the open as the spatially unknown or uncertain (as made addressable by the Ptolemaic grid) on the one hand and the open as the temporary, the provisional (as enabled by the veil) on the other. In Hans Blumenberg's words, the fact that "what is not is equally as real (*intense*) as what is . . . is the precise expression for possibility of the modern will to create overall, for the *terra incognita*, whose untrammeled state invites imaginations."[66] That whatever is not is equally as real as that which is, is the ontological precondition for the existence of the design and for the cultural technique of designing uniting the will to create with the will to conquer.

8

WATERLINES

Striated and Smooth Spaces as Techniques of Ship Design

THE ANTHROPOLOGY OF NAVAL ARCHITECTURE

In 1629 the architect Joseph Furttenbach noted that naval architecture (*architectura navalis*) differed from its civilian and military counterparts (*architectura civilis* and *militaris*) because it takes

> such a defiant and fearful thing as the human heart . . . even further away
> from its natural habitation and the place assigned to it by God, and dares,
> upon the wild and terrible element of the formidable sea, to tame it with a
> wooden, if mightily fortified structure, that it may grant him free residence,
> for year and day, on its round surface and prevailing expanse.[1]

Furttenbach was alluding to the ancient topos of seafaring as a blasphemous endeavor that makes man transgress the boundaries set by gods and nature.[2] With the help of technology humans overcome their own nature as well as divine constraints; they conquer their fear and force the seas to grant them "free residence." Naval architecture therefore has a special relationship to anthropology. By establishing itself alongside *architectura civilis* as the maritime equivalent of fortification architecture, it provides the technological precondition for an anthropological redefinition of man as a cultural rather than natural being that, surpassing animal existence, lives up to its human status by moving from striated to smooth space. The famous speech Ulysses delivers in canto 26 of Dante's *Inferno* upon reaching the Pillars of Hercules (that is, the Straits of Gibraltar), which mark the boundary imposed on humans by gods, is a didactic showpiece of this new anthropology. In "this brief vigil of our senses," Ulysses exhorts his companions, they should "follow the sun," that is, travel from east to west, to "the world without people"—in other words, beyond the boundaries of the *oikumene*.³ Ulysses justifies the hubris of this undertaking by invoking what it means to be human. "Considerate la vostra semenza," consider your sowing, he lectures his crew, "fatti non foste a viver come bruti," you were not made to live like brutes, "ma per seguir virtute e canoscenza."⁴ Humans are cultural rather than natural beings; it is their *curiositas* and their *virtù*, their pursuit of knowledge and audacity, that distinguishes them from animals. Naval architecture thus provides globalization with an anthropological grounding. The idea propagated by natural law that the sea is an open space whose freedom may bring about a global human community (and subsequently global international law and universal human rights) is an aspect of this anthropological justification of Eurocentric globalization. In the context of naval architecture, then, design takes on a significance that it lacked in terrestrial architecture. Designing ships merges with the existential self-design of the modern subject navigating the smooth space of the capitalist flows of goods and colonialist science.

WOODEN LINES

The history of naval design can be divided into three stages. In the first stage, design and construction are not yet separated; the second stage employs Euclidean tools to design ships on paper; and in the third, ship design becomes the domain of experimental sciences that draw on the surrounding medium. In other words, all three stages may be defined in terms of their guiding media: The first centers on wood, the second on paper, the third on water and wind.

This periodization follows a well-established pattern in the histories of science and technology. In the older histories clinging to the notion of a "scientific revolution," it was customary to distinguish between a prescientific age of draftsmanship and a scientific age characterized by the "mathematization" of the mechanical arts.⁵ The visible sign of what David McGee has termed the transition from "craftsmanship to draftsmanship" was said to be the shift toward designing buildings and machines on paper. It is not so much this elegant formula that is in need of correction, but the assumption that the bygone stage of craftsmanship was bereft of all mathematics: that it was a time-, labor-, and material-intensive epoch governed by trial and error and the know-how of experienced master shipwrights. In addition, it is misleading to equate the transition from craftsmanship to "science" with that from the usage of three-dimensional objects in workshops to the focus on two-dimensional objects on paper. Ultimately, "science" is much too loose and imprecise a term to capture the strategies, motives, and technological innovations accompanying the switch to paper-based technical design. It makes more sense to use Bruno Latour's concept of "immutable mobiles," which replaces the macrosignifier "science" with a chain of inconspicuous cultural techniques such as representation, projection, grids, scaling, and so on.[6] Indeed, once we abandon the dramatic scenario of epochal social and economic caesuras in favor of a history of elementary cultural techniques, a new view emerges. Take, for instance, the line: It is no longer the exclusive element of an office-based design method involving dividers, ruler, and paper, but an ontologically "open object"[7] featuring different "agencies" in a variety of historical design cultures. There is neither an ontological nor a historical basis for the claim that it is the exclusive function of the line to represent.

Studies in the history of naval architecture have shown that even prior to paper-based design stages, dockyards were acquainted with design methods employing simple mathematical procedures.[8] Sixteenth-century Portuguese treatises on shipbuilding contain descriptions of geometrical procedures that allowed master shipwrights to determine in advance the rising and narrowing of the bottom of the ship and the tapering off of the hull. The very same techniques can in part already be found in fifteenth-century Italian texts. The earliest known among the latter is a manuscript by Giorgio Trombetta dating from 1445. What Portuguese texts refer to as *meia luna* is there called *mezzaluna* (which is in fact also the name of an instrument). In Portugal this particular procedure was combined with another method subsequently refined in England, in which the master frame—the ship's largest frame—was used as a template for all the others. In England this method was called *whole-moulding*. Moulds were normally made of thin boards that had been nailed together. An older alternative to the *mezzaluna* method was the use of ribbands: One simple method was to mount the master frame midships on the keel and then bend pliable wooden battens, or ribbands, around the master frame and attach them to bow and stern. The shape of the other frames was thus indicated by the ribband;[9] only the specifications of the master frame had to be worked out in advance. The curvature of the keel, in turn, was determined by a slack cord attached at different heights to staves placed at either end of the keel.[10] Rather than by lines on paper or columns of numbers, keel and hull were defined by ropes and elastic bands which, as Tim Ingold has shown in his archeology of the line, are themselves a kind of line. They have surfaces, yet are not drawn on surfaces.[11] It was not the line itself that was the innovation in shipbuilding but its transformation, that is, its connection to a ground.

What characterized the *mezzaluna* method was the use of geometrical aids for determining the rising and narrowing of the hull during the actual construction of the ship in the dockyards. Once the form of the midship or master frame had been established, the form was stored by means of a mould, or template.

Given that the frames were composed of several pieces of wood, each of which had to be cut along the lines of a special mould, several templates were needed. The shipwright then used the same templates for determining the shape of all the other frames. In order to arrive at the desired curvature of the hull, the relative position and inclination of the moulds toward each other were changed. Determining the rising of the bottom of the ship, foreshortening the floor timbers, and securing the correct curvature was not achieved by trial and error, but by means of a geometrical projection procedure called *partisan*.[12] First the entire dimension of the hull (the *compartida*) was drawn as a semicircle (hence the name *mezzaluna*, or half moon). This semicircle was then divided into two quadrants; the quadrants, in turn, were divided into as many equal parts as there were to be frames in the front or back half of the hull. Joining up the matching points resulted in a series of bows. The intersections of chords and radius in turn produced a graduated scale of marks,[13] whose values correspond to the expression $x_i = 1 - \sin \alpha_i$, where α_i is the angle of the radius that touches point i on the quadrant.[14] The scale was engraved, on a scale of 1:1, on a gauge, a piece of squared timber known as a *brusca* in Venice and a *graminho* in Portugal (see Figure 8-1). The gauge was used to determine the diminishing depth, that is, the raising of the bottom, or floor timbers, of succeeding frames.[15] Each mark on the gauge or brusca indicated the increment by which the successive frames (or

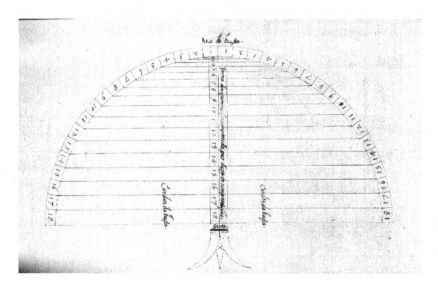

FIGURE 8-1. Fernando Oliveira, drawing of a *graminho*. From Oliveira, *Livro da fábrica das naos*, 93; manuscript, c. 1580. Biblioteca Nacional de Portugal, Lisbon, Autographs Collection.

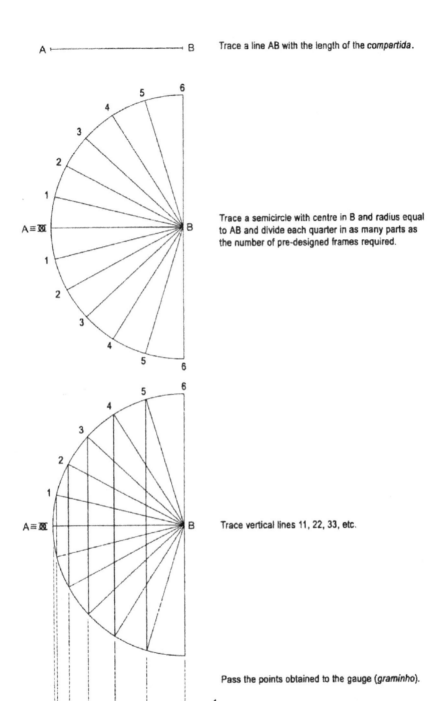

A ├───────────────────┤ B   Trace a line AB with the length of the *compartida*.

Trace a semicircle with centre in B and radius equal
to AB and divide each quarter in as many parts as
the number of pre-designed frames required.

Trace vertical lines 11, 22, 33, etc.

Pass the points obtained to the gauge (*graminho*).

FIGURE 8-2. How the *mezzaluna*, or *meia luna*, was used to mark a scale on the graminho.
© Filipe Castro; reprinted, by kind permission, from Castro, "Rising and Narrowing:
Sixteenth-Century Geometric Algorithms."

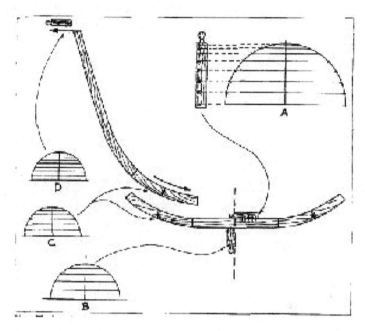

FIGURE 8-3. *Partison* with the aid of the *brusca*. © Sergio Bellabarba; reprinted, by kind permission, from Bellabarba, "The Ancient Methods of Designing Hulls."

their moulds) had to be drawn in, in order to taper the hull (Figures 8-2 and 8-3).

In the wooden world of fifteenth-century arsenals and sixteenth-century Portuguese dockyards, design and construction were no longer identical, as had been the case in the earlier craftsman tradition, because the shape of the frames could now be determined beforehand by geometrical means. Nonetheless it was not yet possible to completely separate design and construction. Using wooden moulds and measuring staffs on a scale of 1:1, the design process took place in the same medium as the construction itself.[16] Indeed, the partially completed vessel itself was "used as an instrument of its own design."[17] The line was both construction aid and part of the construction itself; design as well as construction were carried out in the dockyard. The schooling of apprentices took place in the shape of instructions and advice during the actual work. The master shipwrights "recited aloud the essential details of their projects" in verse, as reflected in the earliest written treatises on shipbuilding, which tend to give an acoustic image of the shipwright's teaching in lyrical form.[18] "Oral and material culture were seamlessly joined."[19]

Once he relocates from the shipyard to his office to conduct his work at the drawing board, the master shipwright transforms into an architect. In other words, this occurs when shipbuilding becomes paperwork.[20] Only then can design turn into an experimenting with forms. The credit for having been the first English shipwright to design ships on paper usually goes to Royal Master Shipwright Mathew Baker. His *Fragments of Ancient English Shipwrightry* (so called by Samuel Pepys),[21] dating from around 1586, contains a picture showing a master shipwright and his assistant at work in a drawing office, with the former wielding outsized dividers over a "plat" of a ship in plan and section (Figure 8-4).[22] The picture tells us a lot about the transformation of ship design into a mathematical art, while also revealing something about Baker's self-fashioning as a mathematical practitioner. The source for a peculiar detail of Baker's representation is the last woodcut in Albrecht Dürer's *Underweysung der Messung* of 1525, which depicts two men constructing a perspectival image (Figure 8-5). The table in Baker's image is a precise copy of the table in Dürer's woodcut.[23] Baker places the shipwright and the master of perspective next to a table on which one can clearly recognize the design of a master frame drawn with the help of a grid. Baker claims Alberti's famous *velum* as a tool of his own, thus intimating a relationship between ship design and *disegno*, the invention of which already in the sixteenth century was (especially by Vasari) attributed to Alberti.[24]

FIGURE 8-4. Mathew Baker, drawing of a master shipwright at work in his office, 1586. Reprinted from Mathew Baker, *Fragments from Ancient Shipwrightry*; manuscript (PL 2820), 1586. © Pepysian Library, Magdalene College, Cambridge.

FIGURE 8-5. Albrecht Dürer, *The Draftsman of the Lute*, 1525. Woodcut, from Dürer, *Underweysung der Messung, mit dem Zirckel und Richtscheyt, in Linien, Ebenen und gantzen Corporen* (Nuremberg, 1525). © Sächsische Landesbibliothek, Staats- und Universitätsbibliothek Dresden; Deutsche Fotothek, Handschriftenabteilung.

Against this background the radical character of paperwork becomes evident: It stands in marked contrast to the templates and measuring staffs of the predesign period. "With Baker," Stephen Johnston notes, "we are now removed from the busy world of the wooden shipyard." That is to say, design now occupies a completely different domain than the actual construction. "Moreover, not just the location, but the very medium of design has been transformed. The master works with dividers on the plan and section of a ship, translating the complex geometry of narrowing into miniaturized form through the technique of scaled drawing."[25]

In addition, switching to paper made experimentation a great deal cheaper. It was not only a convenient medium that enabled Baker to record the midship moulds of other ships; as his *Fragments* shows, it also allowed him to design possible new shapes. The potential of paper rests in its ability to expand the realm of the possible above and beyond the given.

Typically, Baker's midship moulds were drawn over proportional grids and made up of touching arcs of circles. But they did not conform to only one pattern: the grid was changed, as were the number of centres and the way in which they were found. Baker used the pages of the *Fragments* as a medium in which to test new measures and ratios, varying parameters in order to assess their influence on the form of the hull.[26]

The finished drafts found their share of attentive readers among the influential members of the Navy Board and the Privy Council. Indeed, no less a personage than the country's sovereign took an active interest in naval design. In a letter of 1588 (the year the Spanish king sent his Armada into the English Channel) addressed to the three royal shipwrights Peter Pett, Richard Chapman, and Baker himself, the principal naval officers requested that "the Plats [plans, drafts] of the Ships, Galleasses and Crompsters that were lately determined to be built should be set out fair in Plats and brought to my Lord Admiral that her Majesty may see them."[27]

Phineas Pett recounts in his autobiography that in 1612 Prince Henry ordered all master shipwrights to Greenwich: "And every Master Shipwright brought his plats, to the end his Highness might make the better choice for what proportions and kinds of moulds he did best approve of for fitness of service."[28] Plats allowed for the integration of shipbuilding into naval administration. The involvement of the state and its bureaucratic apparatus now went beyond the basic decision whether or not a ship was to be built; it also extended to construction details. It is no coincidence that the beginning of British ship design coincided with the transition of naval politics from irregular privateering to the build-up of a regular navy.

It is interesting to note that the forms of the frames in Baker's plans are drawn through several centers of circles by means of dividers.[29] It was the four-centered arch or Tudor Arch in particular that was used in England as a model for ship frames.[30] Since those arches can be found in such late Gothic churches as the Chapel of King's College in Cambridge (Figure 8-6), galleons are basically capsized churches. Naval architecture is modified church architecture. Sacred buildings are the main reference for nearly everything in traditional architecture.

With Anthony Deane's *Doctrine of Naval Architecture* of 1670, ship design drawings acquired the form that remained standard until the twentieth century. Beginning with Deane, who was a protegé of Samuel Pepys and became master shipwright of the Harwich dock in 1664, the dimensions and the shape of ships were rendered by three plans: the sheer-plan (or plan of elevation), the half-breadths plan, and, most important, the body-plan, from which one could take the form of the frames (Figure 8-7).[31] While the vessels increased in size, this method of ship design stayed remarkably constant over the centuries.

But as late as 1625 it was still one of the tasks of the master shipwright to fabricate moulds: "I made moulds and sent them into the woods by one Thomas Williams, shipwright," Phineas Pett notes in his autobiography.[32] The shipwrights ventured out into the woods in search of trees that could be cut into fitting shapes. Much as a master painter chose the right wood for his panels, the shipwright carefully selected the trees that would provide parts of the hull. Plats in hand, they combed the woods to find trunks and branches with the curvature required by the design.

FIGURE 8-6. King's College Chapel, Cambridge. Engraving, from David Loggan, *Cantabrigia Illustrata* (Cambridge, 1690). Reprinted by kind permission of St. John's College, Cambridge.

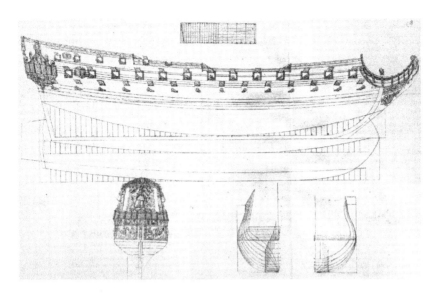

FIGURE 8-7. Anthony Deane, "The ship's draught completed in every part." Drawing, from Anthony Deane, *Doctrine of Naval Architecture* (1670). Reprinted from *Deane's Doctrine of Naval Architecture, 1670*, ed. Brian Lavery, 70.

WATERLINES

The line that relates the form of the ship to the surrounding medium is the waterline. Running along the hull, it intersects all frame sections. Shipwrights in the times of Baker or Deane could tell from experience the draft of empty, half-empty, and fully loaded hulls, which allowed them to draw three water-lines onto their construction plans. During the second half of the eighteenth century the semantics of this line were radically reinterpreted. It was de-referentialized and transformed into a sign that no longer described the form of the hull with help of the tools of striated space, but with the help of smooth space itself.

In 1738 the French geographer Philippe Buache became interested in map-ping the bottom of the sea. In 1752 he published the first map depicting the shape of the sea bottom between France and Britain. Buache was probably the first cartographer to use isometric lines that show depth below a water-level plane of reference, so-called isobaths, in the open sea.[33] Although he was most likely not the first to represent the maritime depths, the publication of his map and the memorandum explaining his methods in the *Mémoires* of the French Acad-emy of Science attracted widespread attention. The Dépot des Cartes et Plans de la Marine started to evaluate hundreds of logbooks into which over several centuries captains had entered soundings carried out near the coast.[34] One could easily copy the results of these soundings onto a map; and if the numbers were sufficient, it was possible to connect them and arrive at a topographical representation of the sea bottom by means of contour lines. A pioneer of this approach was Marcellin Du Carla, a geographer from the Languedoc. In 1782 the mapmaker and publisher Jean-Louis Dupain-Triel printed Du Carla's manuscript, entitled *Expression des nivellements, ou Méthode nouvelle pour marquer rigoureusement sur les cartes terrestres et marines les hauteurs et les configurations des terreins,* which contained the first complete description of the method of contour lines (Figure 8-8).[35] To explain the underlying concept of contour lines, Du Carla reminded his readers of a phenomenon familiar to anyone who has ever walked along a beach: they were to imagine an all-embracing deluge that rose and fell in degrees, as a result of which the surface of the water would leave

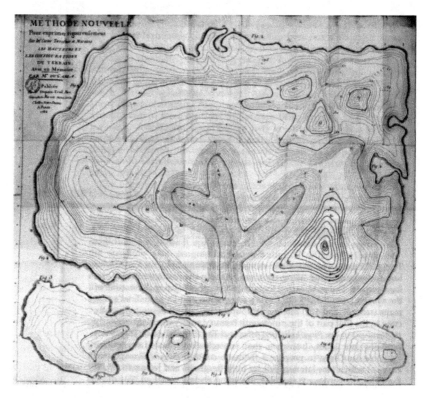

FIGURE 8-8. Marcellin Du Carla, hypothetical contour map, 1782. Engraving, reprinted from Du Carla, *Expression des nivellements, ou Méthode nouvelle pour marquer rigoureusement sur les cartes terrestres et marines les hauteurs et les configurations des terreins* (Paris: L. Cellot, 1782).

behind shifting lines in the sand. Contour lines, in other words, were analogous to the high and low water lines the ocean leaves behind on the beach.[36] The sea bottom is not without form, but its form is a "shapeless form."[37]

Yet before contour lines could be introduced as a new technique into ship design, bathymetry (the technique of surveying the sea bottom) had to be transferred onto dry land. The basic idea behind this transfer was already expressed by Du Carla: Contour lines on a map, he wrote, would enable the reader to check what one could or could not see from a given point. A certain group of engineers has a particular interest in such data: those specializing in fortifications. After all, the most important aspect to keep in mind when designing a new fortification was to place it in such a way that it would not possible for enemy cannons to fire into it from any given point in its surroundings. Hence, the fortification engineers trained in Gaspard Monge's school in Mézières seized upon Du Carla's method and further developed it by transferring it onto land. The magnitudes of elevation they recorded on their maps were in fact magnitudes of depths measured downwards from an imaginary horizontal plane that intersected the highest point in the surrounding area (which was given the value o).[38]

Assembled completely from Euclidean bodies, fortifications were analyzed as if a second deluge had taken place. Everything is forever under water, which renders it possible to describe everything as *height below*—that is, as height below the surface of the water. Euclidean bodies become special cases of irregular bodies, since the method of contour lines derived from a method for surveying irregular, shapeless sea bottoms. While Du Carla seized upon the idea that the lines on the shore left by the sea at the highest tide level could be generalized as contour lines, one of the most influential naval architects of the late eighteenth century, Henri-Louis Duhamel de Monceau, General Inspector of the French navy, arrived at a similar idea, namely, that a ship's waterlines, which since the days of Baker and Deane had been no more than marginal marks on the design plans, could be generalized as contour lines as well. In 1752 Duhamel de Monceau published his *Élemens de l'Architecture navale*, a book that dealt primarily with the drafting methods in ship architecture. In 1771 Marcellin Du Carla met Duhamel de Monceau, Jean-Paul Grandjean de Fouchy, and Philippe Buache in Paris.[39] In the second edition of his book Duhamel explicitly performs the transformation of the waterline from a referential sign to an abstract line of contour.

> You will now understand that, if one unloads a ship step by step, one can draw as many waterlines on its hull as one wishes, provided that one pays attention to keep it in one and the same position. From which fact one can

easily deduce that all parallels to the surface of the water that one draws on the hull will be waterlines . . . waterlines which may be multiplied *ad libitum*.[40]

Duhamel advised that plans be drafted that define ships by means of imaginary waterlines modeled on the contour lines used to represent the shapeless forms of the sea bottom. One indication that Duhamel adopted Du Carla's idea is that both speak of the autography—literally, the self-writing—of the water on the boundary between the finite and the infinite.

Toward the end of the eighteenth century the encounter of maritime surveyors and naval architects gave rise to the idea that waterlines on plans did not need to represent real waterlines. This, in turn, implied another idea: water itself could be used to mark the contours of a hull. In fact, the latter could be generated by the intersecting line between the solid body and the liquid medium, thereby rendering obsolete the construction of vault arches with the help of dividers (Figure 8-9). The intersections of the frames may thus be interpreted as waterlines resulting from the piecewise vertical immersion of the hull in water. The method of imaginary planes opened the way for describing any body as a streamlined body on paper (or in the computer). As epistemic things, then, ships were submarines even before the latter had been invented. All ships emerge from under water. This was a completely new way of defining bodies. While Baker's design method placed the Euclidean dividers in the center (with a subtle reference to Dürer's apparatus of perspective), the waterline method is more closely related to the domain of non-Euclidean bodies. Shape emerges from

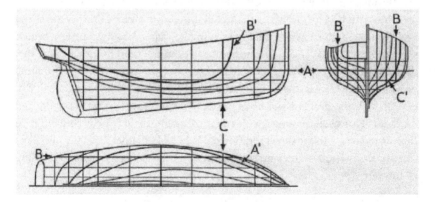

FIGURE 8-9. A set of lines used to record the shape of a pilot boat, c. 1900. Each set of curves appears in the other two views as a series of straight lines. Reprinted from Booker, *A History of Engineering Drawing*, 71.

shapelessness. It is no coincidence that the origin of this new conceptualization is closely linked to the old biblical view of the sea bottom filled with debris from the original *tohuwabohu*—a fearful, misshapen, and monstrous landscape.[41]

If according to Euclidean or scientific conceptualizations of striated space the line limits or excludes white space, the science of smooth space conceptualizes this white interim space as made up of an infinite number of possible mental lines. Each line is the actualization of a virtual form contained on the white ground itself. It is hardly accidental that this new definition of a new, as it were, shapeless hull emerged simultaneously with the first experimental methods of ship design. Prior to 1685, the aforementioned Anthony Deane had carried out a series of experiments together with the military engineer Henry Sheeres. The two compared the speed of planks cut in the shape of the waterlines of seven different ships. Each plank was pulled through a 25-meter trough by a falling weight and timed by a pendulum. It appeared that the fastest ship was the obsolete Venetian galley, while the slowest was a Dutch vessel.[42] In the eighteenth century other naval architects performed "flotation tests" using simple shapes. In the early 1750s a constructor named Bird carried out similar experiments with different hull shapes in a 10-meter tank.

In 1775 the Swede Frederik Henrik af Chapman published his *Tractat om Skepps-Byggeriet (Treatise on Shipbuilding)*, which was translated into English in 1820.[43] His *Architectura Navalis Mercatoria* of 1768 had been the first treatise by a naval architect to draw on design methods other than the half moon and the dividers. As part of his hydrodynamical experiments, Chapman had seven different hull shapes, each approximately 70 centimeters long, pulled through a 20-meter tank. Chapman was the first since Leonardo da Vinci to show an image of the streamlines produced around a body in motion (Figure 8-10).[44] If the Phoenician master shipwright Tridon in Paul Valéry's dialogue "Eupalinos ou l'Architecte" "thought a ship should almost be fashioned by the very wave itself!"[45] then such a thought was the result of a new experimental science, fluid mechanics or hydrodynamics, that conducted experiments on "the resistance which a ship in motion meets with from water" to determine the optimally streamlined design.[46]

In France, experiments on the resistance of fluids and tests of streamlined hulls were first carried out by a navy lieutenant and member of the *corps de génie*, Jean-Charles de Borda, in 1767 and 1769. In 1768 he met the captain of the port of Lorient, the Comte de Thévenard. Thévenard succeeded in raising funds from the navy and the French East India Company to excavate the largest testing basin ever constructed until 1870, a canal 70 meters long and about 4 meters in width, to carry out large-scale experiments towing solids at the surface

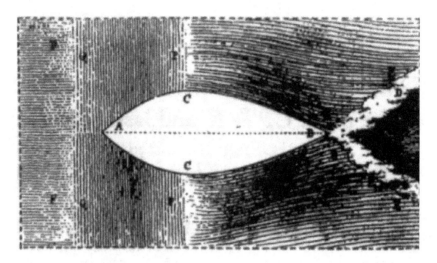

FIGURE 8-10. Fredrik Henrik af Chapman, drawing of streamlining produced around a body in motion, from *Tractat om Skepps-Byggerie* (Stockholm, 1775). Reprinted from Chapman, *A Treatise on Shipbuilding, with Explanations and Demonstrations Respecting the Architectura Navalis Mercatoria,* fig. 14.

and at varying depths.[47] Like Du Carla and Buache, Borda belonged to the group that met Duhamel de Monceau in Paris.

TRITONIC CULTURE

1932: Norman Bel Geddes designs a ship which at first glance still retains features of Valéry's *objet ambigu*. One can see that it is a manmade machine, yet it nonetheless has taken on the appearance of a thing shaped by wind and water, like a smoothly polished bone (Figure 8-11). Is it supposed to move on, above, or under water? If a model of Bel Geddes' ocean liner had been washed ashore at the feet of Valéry's Socrates, he most likely would not have been able to make heads or tails of it. Everything has been designed to offer minimum resistance to wind and water. All projections including smokestacks and lifeboats are enclosed within a streamlined shell. "The entire superstructure is streamlined. Every air pocket of any kind whatsoever has been eliminated. . . . The single protrusion is the navigator's bridge and this is cantilevered and similar in shape to a monoplane wing, consequently offering a minimum of resistance to the air."[48] Bel Geddes himself realized that his ocean liner had turned into something ships, as epistemic objects, had been ever since their design came to depend on imaginary planes and contour lines: "[O]n stormy days the ship is as

FIGURE 8-11. Norman Bel Geddes's streamlined ocean liner model. Photograph by
Maurice Goldberg, c. 1932. Harry Ransom Center, The University of Texas at Austin.
Image courtesy of the Edith Lutyens and Norman Bel Geddes Foundation.

enclosed as a submarine. . . ."[49] Now, Natasha Pulitzer Los wrote, as the sea had
lost its terror, ships of the interwar period started to change their meaning. This
implied, however, that the house was changing its meaning too, "[f]rom the
Machine to Live In to the House to Sail In."[50] As R. Buckminster Fuller noted
in the margins of his design of the Dymaxion House of 1929, "houses may be
considered aerodynamically as little ships whose standard cruising is 12 miles
an hour."[51] While around 1600 the laws of hull construction were deduced from
the nave and the design of the ship's superstructures from land-based fortifications, Buckminster Fuller in the twentieth century derives the laws of the house
from the ship. Architecture underwent a radical turn from land to sea. Hydrodynamics or fluid mechanics enabled a non-Platonic cultural technique of design. Tools for the description of the irregular, the shapeless, and the
non-Euclidean became design tools for ships and houses, whose architecture
is no longer based on Euclidean geometry. A Euclidean shape such as the cube
is just a special case of the non-Euclidean, as rational numbers are simply a special case of real numbers, and the house is no more than a special case of the
ship. The "Nomos of the Earth" in the twentieth century is ruled by Tritons[52]—
demiurges of the depths who circumscribe solid bodies by contours and stream-
line and shape a body by means of the controllable models of the ocean itself.
Houses, Adorno concluded in 1944, belong to the past.[53] A dwelling that does
justice to the existential situation of the present is located somewhere between
the extremes of container and floating palace.

9

FIGURES OF SELF-REFERENCE

*A Media Genealogy of the Trompe-l'oeil in
Seventeenth-Century Dutch Still Life*

As opposed to natural things embedded in a surrounding landscape, the flowers, fruits, pastry, fish, meat, cheese, or game we encounter in still lifes are objects of domestic use. They appear as things that by definition refer to absent human actors. In Heidegger's parlance, still-life things are equipment (*Zeug*).[1] Victual still lifes refer to actions either past, pending, interrupted, or (for representational reasons) deferred. The things are no longer ready-to-hand (*zuhanden*); instead they are, in recalcitrant fashion, merely present-at-hand (*vorhanden*).

It is this recalcitrance, resulting from the interruption of the referential context of the "equipment," that endows still-life things with their aesthetic charm and inherent power. Removed from their context, they appear to exist in a highly artificial tranquility which, in turn, abandons them to their specific forms of decay and perishability. They attract, as both witness and sign of this decay, all kinds of flies, beetles, and other vermin. But then again, the disruption of the referential context is also an appeal to take possession of these things. Not only do they invite in-depth scrutiny, especially when revealing their innards; they also appeal to the observer to anticipate and complete the interrupted actions. Still-life objects are action-related hypotheses. They involve the observer in the painted worlds in ways that transcend the cultural techniques of aesthetic perception: They urge us to mistake the representation of the object for the object itself. Yet although this particular confusion of art and reality, which goes back to Pliny's endlessly recycled anecdote of the contest between Zeuxis and Parrhasius, is part of the standard seventeenth-century praise of artistic accomplishment,[2] the confusion brought about by still lifes reaches deeper than the rhetorics of the genre. It is the *appetitus* rather than the contemplation mobilized by the food still lifes that enjoyed such great popularity during the baroque era, for it is the *appetitus* that mistakes representation for object. Just as the ubiquitous flies are attracted by the painted fruits, our gaze is transfixed by what our gluttony takes for the real thing. As action-related hypotheses able to fool observers into confusing representation for reality, still-life objects exert a strong influence that as late as the seventeenth century was registered as a physical effect. The notary and history painter Cornelis de Bie wrote of the impact of the famous still lifes by Jan Davidszoon de Heem:

JAN DE HEEM
FRUIT PAINTER FROM UTRECHT

Ye women who are getting fatter, do not look so intently
at the painted fruit that seems to resemble life,
lest your foolish eyes torment your heart,
your unborn child be affected.
The sight of such art will awaken sensual pleasure.
When you stalk the works of Jan de Heem
what he can so sweetly depict on smooth panels
may tickle the heart of a pregnant woman.
It is wondrous what he can achieve with such miracles
that reveal in fruits the true appearance of life.
So sweet his cherries, so ripe the muscadine grape,
the pomegranate, peach, and apricot, all laid out there
as natural as if it had grown:
with such precision does he make this art match real life.[3]

The painted object that more than any other explicitly prompts observers to mistake a representation of the thing for the thing itself is the trompe-l'oeil; and more than any other genre, still life exhibits a penchant for the trompe-l'oeil. This inclination may on occasion result in extreme cases in which the trompe-l'oeil comes to dominate the entire painting, as in the case of Cornelis Gijsbrecht's quodlibets and tackboards, not to mention his *Reverse Side of a Painting*.[4] But leaving aside these extreme cases, the still life possesses genre-specific features that predestine it for trompe-l'oeil details and similar effects, so many of which are to be found, more or less conspicuously, in Dutch still lifes.

What I am proposing here is to read the trompe-l'oeil as an instance of self-reference of a hybrid text-image medium. The medium in question is the illuminated manuscript page of the late fifteenth and early sixteenth century. Its differentiation, I will argue, gave rise to Dutch still life, and in the course of this differentiation the trompe-l'oeil emerged as a sign that pays homage to the reflexivity of the two-dimensional materiality of the illuminated manuscript page. This reflexivity, in turn, is expressed by addressing and observing the unity of the difference between the imaginary pictorial space and the real space of the reader/observer.

TRANSPARENCY AND OPACITY

Louis Marin has argued that with the trompe-l'oeil, mimesis, as it were, overshoots its own target. If according to Alberti's famous metaphor the space caught within the rectangle of a painting is an open window onto the world (transparency), then, Marin adds, the trompe-l'oeil effectively closes this very window (opacity). Marin refers to this as "excessive mimesis." Following the historian Pierre Charpentrat, Marin denies that the trompe-l'oeil and other forms of excessive mimesis are still part of the order of representation. Rather, they belong to the order of hallucination.

> Perhaps it would be useful to reflect here on relief in the trompe-l'oeil object. For in most cases, its relief—its third dimension, whose production in the surface of the canvas is the triumphal accomplishment of mimesis— plays its useless and excessive role in a very different way. . . . The "thing" stands out on the canvas; the apparition comes to meet the gaze; the double is strangely detached from the surface of the plastic screen. Strangely, not completely: the hallucination is in the process of being born. Let us consider for a moment a still life in which a fly has landed, or one in which a drop of water is flowing, limpid as a tear. Where has the fly landed? Where is the drop flowing? On the corolla of the flower, on the side of the flask, *represented*? A strange hesitation, for, to my naive and attentive gaze, which limits itself strictly to seeing . . . here is the fly poised not on the petal but on the surface of the canvas, here is the drop gliding not down the crystal of the carafe but down the diaphanous plane of the *tavola*.[5]

The trompe-l'oeil effect occurs when the frame or part of it is doubled by a painting—or in the extreme case, when the real frame is replace by a painted one. The pure trompe-l'oeil is characterized by a fake vertical support and the absence of both a horizon and spatial depth. These are generic markers of the still life. However, unlike Baudrillard, who was interested in isolating the pure trompe-l'oeil,[6] we are here first and foremost concerned with investigating impure and hybrid forms, including still lifes. Still lifes appear to be closely related to the trompe-l'oeil phenomenon because their lower edge is normally occupied by a tabletop that in many cases reaches the intersection of the visual pyramid, that is, whose front edge touches the imaginary plane that coincides with the real material frame (be it canvas, copperplate, or wooden tablet panel). Anything reaching beyond this edge appears to protrude into the real space of the observer, located in front of the painting (in Dutch still lifes the objects most likely to do this are knife handles, which appear to protrude into the observer's space, much like a handle attached to the painting itself; the ubiquitous plate edges; or the inevitable lemon peel dangling off the table). It is this particular edge—the table's edge that doubles as the lower edge of the still life—that is in need of both art-historical and media-theoretical scrutiny. It is a sign carrier contained within the painted object, the *mise-en-abyme* of a medium within another medium.

In other words, the trompe-l'oeil is neither mannerist embellishment nor the result of an extreme outgrowth of baroque still life (as in the case of Gijsbrecht). On the contrary: Genealogically it is impossible to separate still lifes from trompe-l'oeil. The latter is neither an accessory nor an ingredient added to the "real" still life in order to increase its illusionary effects. In the following I will try to invert this standard perception. Baroque still lifes testify to a taming of the trompe-l'oeil. They point toward a constantly attempted yet never successfully completed process of avoiding the trompe-l'oeil.

JAN VAN KESSEL: THE ART OF AVOIDING THE TROMPE-L'OEIL

The following analysis is based on the assumption that still-life trompe-l'oeils retain a form of self-reference that can be traced back to the development of the illuminated manuscript page in the fifteenth-century Netherlands. To understand this development, it is necessary to set aside the prominent (and somewhat pompous) baroque still lifes and instead focus on deviations and variations located at the margins of the genre. One such marginal variation is Jan van Kessel, a minor player in the history of Dutch still-life painting, yet one whose work exhibits a set of features that may be deciphered as an ongoing effect of the illuminated page. He was a grandson of Jan Brueghel the Elder, a.k.a. "Flower Brueghel," a frequent collaborator of Rubens. Van Kessel's *Insects on a Stone Slab* (Figure 9-1) demonstrates the residual persistence of the illuminated manuscript page in a different medium. A stone slab is covered by a number of beetles and other insects indigenous to South America. Some are perched on the front of the slab, others are crawling over and along its side and top edge. A second slab, jutting out toward the observer, intersects the first at more or less a right angle. The background is a landscape at dusk with a fortress in the distance. At first glance there is little that is remarkable about this painting. Insects abound in numerous Dutch hunting, kitchen, flower, and fruit still lifes. Attracted by the putrid odors of fruit and decaying game, they are unmistakable vanitas

FIGURE 9-1. Jan van Kessel, *Insects on a Stone Slab*, undated. Oil on copper, 17.5 × 23.5 cm. © Kunstmuseum Basel. Photo: Martin P. Bühler.

symbols. But in addition to transmitting the obligatory reminder of death and transience, they time and again act as operators of an optical illusion. Frequently depicted larger than they should be relative to the other elements in the painting, they tend to be placed where they create a trompe-l'oeil effect— astride the curvature of a piece of fruit, atop a rooster's crest, or on the painted frame of the painting. In the case of *Insects on a Stone Slab*, however, the insects have become the principal actors. The still life is "gone," as it were; the trompe-l'oeils alone remain.

On closer inspection this is in many ways an odd painting. First, even if we assume that some of the South American beetles are fairly large, the height of the slab, reminiscent of a towering stone monument, can hardly exceed three to four inches, which would imply that the artist or observer is lying flat on the ground in front of it. If the slab were as large as its design indicates, the airborne insects would be as big as World War I biplanes. A second peculiar feature is the faulty perspective used to render the second slab. Although it intersects the first at a right angle, the insect perched on it is almost on the same plane as the beetle on the other slab; it is as if it were resisting the artistic dictates of central perspective. As we shall see, this minor discrepancy points at nothing less than one of the central issues of Dutch still life.

Third, the slab sports a jointed frame reminiscent of a mural, which in this natural surrounding serves to make it seem even more out of place. Fourth, the naive rendition of nature clashes with the meticulous depiction of the insects that appear to have been drawn with a much finer brush. As a result it is not clear how one is supposed to look at this painting. Insects, slabs, and the surrounding landscape do not form a continuum; each occupies its own world.

A second example: Take a look at Jan van Kessel's *Insects and Fruit* (Figure 9-2). White currants, moths, a caterpillar, mosquitoes, and a mix of other insects are spread across a ground of opaque white. Once again, the observer is confounded by an ambiguous surface. Judging by their shadows, some of the insects sit on that ground as if it were a horizontal plane extending backwards into space, while others appear to be using it as a vertical wall. Either van Kessel's space is warped, or some of the insects are inhabiting the imaginary space *within* the painting while others are sitting *on* the real picture.

There is a simple way to explain this ambivalent space. As Svetlana Alpers and others have argued, we may be dealing with the artistic outcrop of a scientific dispositive of visualization and description that aims to present to voracious eyes different vantage points on items of nature normally removed from sight. Such a descriptive and explorative gaze, fueled by a "microscopic taste for displaying multiple surfaces,"[7] strives to open up things for inspection as if they were little machines. This opening-up corresponds to the late-seventeenth-century practice of presenting still-life objects such as cheese, pies, lemons, fish, or nuts in a damaged state—sliced, cut up, partly eaten, or burst—revealing their inside. The primary goal is to depict the composition and functioning of objects; the secondary concerns their position in spaces that vary according to the desired view, as in the case of technical drawings that on one and the same page show an object from above, in profile, and as a cross section. This is a persuasive explanation, no doubt, yet the question remains why van Kessel offers no clear intradiegetic disambiguation to clarify his ambivalent space. It would have been easy to achieve this by means of the niche architecture common to early still lifes. The alternate explanation I would like to submit comes down to viewing this space as the result of a conflict between two cultural techniques—*gazing* and *reading*. At one point these two techniques were interwoven, but in the course of medial differentiation the techniques themselves were differentiated. Trying to navigate van Kessel's ambivalent spaces, then, we are confronted with an incomplete, only partly successful media-technological differentiation. The older cultural technique, which had subjected both writing and image to a common figurative meaning, is not simply replaced by the "modern," disjunctive techniques of viewing images on the one hand and reading texts on the other; it is marginalized and moved into a diaphanous zone.

FIGURE 9-2. Jan van Kessel, *Insects and Fruit,* c. 1655. Oil on copper, 11×15 cm. © Rijksmuseum, Amsterdam.

What, then, is happening in *Insects on a Stone Slab?* It appears that van Kessel tried to add an intradiegetic vertical carrier and a horizontal dimension to his insects in order to disambiguate the ambivalent surface of the copper plate. However, in doing so he did not wholly move beyond the cultural techniques of reading images linked to the quodlibet's image carriers. In other words, the *mise-en-abyme* of *Insects on a Stone Slab* constitutes an attempt to *avoid* a trompe-l'oeil effect by means of metapainting. The copper plate ended up as a stone slab, and what we see is a van Kessel within a van Kessel. The stone slab is a compromise between readability and visibility.

JORIS HOEFNAGEL: METAMORPHOSES OF FIGURE AND GROUND

There is a well-known model for van Kessel's quodlibets, in particular for his insect studies, which belongs to the realm of the curiosity cabinet as well to the final blooming of Flemish book art: Jacob Hoefnagel's 1592 *Archetypa studiaque patris Georgii Hoefnagelii.* As indicated by the title, it is an emblematically enriched sample book of copper engravings for artists and other *philomusis* ("Friends of the Muses") based on watercolor paintings by his father Joris (Georg). Hoefnagel's meticulously rendered insects, snails, flowers, and fruits were part of Rudolf II's famous cabinet of curiosities. It is known that the *Archetypa* served as a model for many other artists; its influence in the still-life genre reaches as far as the late seventeenth century.[8]

The Brukenthal Museum in Hermannstadt, Romania owns a watercolor painted on parchment by Joris Hoefnagel that sheds light on the trompe-l'oeil context and constitutes an important link in the chain of metamorphoses connecting the art of illuminating manuscripts to Dutch still-life painting (Figure 9-3). The most revealing detail for the intermedial reappearance of the manuscript page as a painted object is the frame: The parchment features a painted wooden frame whose lower edge is ornamentally adorned, ending on both sides in a squiggle. The picture has received a trompe-l'oeil frame. On both sides the frame is equipped with small eyelets and fixing screws into which roses have been stuck. These roses possess a hybrid, metamorphic dimensionality. Sticking out from the trompe-l'oeil frame, their stems appear three dimensional, while their blossoms share the bidimensionality of the parchment surface. The left, faintly rendered rose in particular does not appear to be lying *on* the surface or sticking to the frame, but is content to appear as what it is: a drawing on parchment that (unlike the split peach in the lower right) casts no shadow. Further revealing details are found on other miniatures, for instance, on a 1589 watercolor that has been called the first known independent still life by a Netherlandish artist (Figure 9-4). It dates from the period when Hoefnagel

FIGURE 9-3. Joris Hoefnagel, *Miniature with Snail*, c. 1590. Watercolor and gouache on vellum, 22 × 33 cm. © National Brukenthal Museum, Sibiu, Romania.

FIGURE 9-4. Joris Hoefnagel, *Still Life with Flowers, a Snail, and Insects*, 1589. Watercolor, gouache, and shell gold on vellum, 11.7 × 9.3 cm. Metropolitan Museum of Art, New York. © bpk / The Metropolitan Museum of Art.

worked in Munich as a miniature painter at the court of Duke Albrecht V of Bavaria. The flower vase, the prototypical still-life object, rests on a horizontal console that recedes back into space and is connected to a vertical wooden trompe-l'oeil frame which, in turn, is connected to the actual frame. This horizontal board is visually at odds with the painting's surface—it is difficult to say on which level it is located. The same applies to the vase, which resides in an impossible space between picture frame and vellum.

How are we to explain this visual paradox? One clue can be found in another work by Joris Hoefnagel. Continuing the tradition of Flemish manuscript illustration, Hoefnagel was the last of the great illuminators of the Habsburg Empire. In the 1590s he was commissioned by Emperor Rudolf II to illuminate Georg Bocskay's *Mira calligraphiae monumenta*, a model book of calligraphy, by adding fruits, insects, and flowers to almost every page. The composition of certain pages provides a decisive media-historical clue regarding the origin of the protruding semi-two-dimensional and semi-three-dimensional objects we encounter in the watercolors and the *Archetypa* models. Folio 37 depicts a nonesuch (an exotic flower also known as Maltese cross), a mussel, and a ladybug (Figure 9-5).[9] At the upper left edge the curlicues of the initial letter expand into an ornamental pattern. This explains the ornate shape of the wooden frames on the *Archetypa* pages: It sprang from writing. It is a calligraphic ornament that has attained object status. In the lower part, however, there is a tell-tale detail that conjures up the trompe-l'oeil effect and sheds light on the link between the objectification of writing and the ambiguity of the surface. A long line snakes its way from left to right across the page, but before it ends in a flourish it crosses a little gangplank, as it were, formed by a slit cut out of the vellum of the page into which the stem of the flower has been stuck. This slit appears to turn the two-dimensional page of the book into a three-dimensional object. The two-dimensional writing surface is transformed into an illusionary three-dimensional object that, paradoxically, appears to be resting on itself. The vellum, that is, the carrier itself, into which the line has been inscribed, becomes a trompe-l'oeil: the image carrier steps out of itself to become an image object. Above the slit the flower is a two-dimensional iconic sign, but below it has turned into a three-dimensional trompe-l'oeil object casting a shadow on the vellum with which it no longer seems connected.

However, the full extent of this hallucinatory excess with which Hoefnagel challenges the bidimensionality of writing only becomes evident once we turn the page and inspect the verso (Figure 9-6). The color and ink of the recto can be seen through the vellum: pale mirror images of the text, the flower, the ladybug, and the mussel. These, of course, are not drawings but real transparency effects. But in the middle of the page we see the stem of the nonesuch that appears to pierce the page and lie on the narrow vellum strip. The shadow of the strip as well as the dark edges of the hole and the small "visible" piece of the stem are the only elements that have been painted on this side.[10] By means of this gesture, the illuminator triumphs over the calligrapher; image vanquishes writing. Hoefnagel transforms the page itself into an object whose topology

FIGURE 9-5. Joris Hoefnagel, illumination of a page (folio 37r) of *Mira calligraphiae monumenta* (Vienna, 1591–96), by Georg Bocskay. Tempera, gold, silver, and ink on parchment and paper. The Getty Center, Los Angeles. Reprinted from Hendrix and Vignau-Wilberg, eds., *Mira calligraphiae monumenta: A Sixteenth-Century Calligraphic Manuscript.*

FIGURE 9-6. Joris Hoefnagel, illumination of verso of a page (folio 37v) of *Mira calligraphiae monumenta*. The Getty Center, Los Angeles. Reprinted from Hendrix and Vignau-Wilberg, *Mira calligraphiae monumenta: A Sixteenth-Century Calligraphic Manuscript*.

oscillates between bi- and tridimensionality, and that by the act of turning the page attains real tridimensionality. Not only does the trompe-l'oeil refer to the vellum as the real image carrier (as also happens in the case of trompe-l'oeil insects on still lifes or on van Kessel's quodlibets), but the very act of turning the page folds the illusion of the three-dimensional stem into the real tridimensionality of space.[11] Here, the trompe-l'oeil invades the space of the observer in a real rather than merely illusory manner. The play of recto and verso enabled by objects such as book pages that can be turned creates an ambivalent threshold zone between the imaginary of the image and the real of the reader/observer. The medial conditions for the birth of a hallucination have been met—a hallucination that is at its very core a medial operation.

REORGANIZATION AND MEDIAL SELF-REFLECTION OF THE ILLUMINATED MANUSCRIPT PAGE AROUND 1480

The claim that the origin of these trompe-l'oeils, that is, the medial basis of such a mixture of varying categories of being (image carriers and objects), can be found in fifteenth-century Dutch book illumination is not new. It has already been made by book specialists and Hoefnagel experts.[12] There is no doubt that Hoefnagel's hallucinatory trompe-l'oeil art is directly linked to the famous art of Flemish manuscript illuminators. Hoefnagel himself illustrated the Book of Hours of Philipp of Cleves.[13] At its margins we find penetrations just like those of the *Mira calligraphiae monumenta*. Folio 57 shows a columbine that appears to be fastened to the parchment by a needle, and upon turning the page we find a needle piercing the page. Needless to say, columbine and needle allow for iconographic interpretations. The columbines of the van Eycks' *Ghent Altarpiece* or Hugo van der Goes's *Portinari Altarpiece* refer to Christ or represent an appeal to God, while the needle alludes to the nails of crucifixion. The trompe-l'oeil is thus endowed with a dimension of meaning that relates the emergence from the second into the third dimension to the body of Christ and its presence in the Eucharist.

The borders of the Hours of Mary of Burgundy (Vienna codex 1857) could well have been a model for the interplay of figure and ground in Hoefnagel's illumination of folio 37 of the *Mira calligraphiae monumenta*, but it is more likely only a persuasive piece of evidence for the close proximity between Hoefnagel and late fifteenth-century Flemish book illumination. The extended descender of the letter i leading off *illud* in the bottom line of the text is transformed into an ornate line ending in a spiraling tendril. Three times the ornamental line is threaded through, as it were, the illusionary double holes penetrating the parchment that—just as in folio 37 of the *Mira calligraphiae monumenta*—appears to form a little crossing. As with Hoefnagel, a line leads across the flap, and once again the flap turns the two-dimensional page into a three-dimensional object, although in this case the line itself does not transform into a flower stem or any other three-dimensional object. The line is a line and yet it is not a line. It starts out as a piece of writing and ends as an ornament, moving in inexplicable fashion across, through, behind, once more through, and then again across the front of the page.

Stylistically and with regard to certain details, Hoefnagel's work shares marked similarities with the so-called Ghent-Bruges school of manuscript illuminators that flourished between the late fifteenth and early sixteenth century. What ultimately caused Hoefnaegel's virtuoso self-reflection of the book page in the sixteenth century, however, is a change that radically affected the appearance of Flemish hour and prayer books in the late 1470s. In 1947 Otto Pächt famously attributed this reorganization of the illuminated page exclusively to the Master of Mary of Burgundy:

> He radically changed the relationship of both miniature and border to the page. In the border-zone the illusion is created that the branches, flowers, insects and birds etc. have been dropped on the page, quite casually, loosely scattered, casting their shadow on the coloured foil.[14]

As Pächt explains, two kinds of illusions are enacted on the page: that of depth and that of a picture frame, with the latter making use of trompe-l'oeils. "The plane of the page, however, is conceived of as a barrier dividing the two kinds of space, the imaginary space of the picture behind the page and the space of reality in front of it." And while the border ornament composed of trompe-l'oeils "has moved closer to the spectator, the scene in the picture has receded further into the background."[15] Because one and the same perspective is employed for both border and miniature, the latter turns into an illusionary niche. Today we view this development in a more differentiated way, just as the identity of the Master of Mary of Burgundy has in part been called into question.[16]

The restructuring of the book page after 1480 may be understood as resulting from the self-reflection of the illuminated page. It has been attributed to a "new spatial thinking,"[17] which reputedly first affected individual pictorial components and then moved toward integrating a third dimension. But this explanation merely serves to create something in even greater need of explanation: the proposed "new spatial thinking." However, once we set aside this nebulous concept and instead focus on the empirically accessible features of intermedial competition, we can trace how, faced with the double challenge of panel painting and the printed book, the manuscript page starts to reflect, first, on its own materiality as a flat object comprised of heterogeneous elements (text, miniatures, initials, borders), and second, on the devotional practices essential for its survival.[18] As a result of this medial self-reflection, the border zone turns into a threshold that encompasses both the real space of the observer and the imaginary space of the miniature. Flowers start to grow out of the book page; arriving insects add a further degree of reality. The trompe-l'oeil, then, is a second-order result of the process in the course of which the book page becomes aware of its own materiality as a two-dimensional object that can be turned around, and then proceeds to medially thematize its materiality.

One example of the trompe-l'oeil as an image resulting from medial self-reflection may be found in the Grimani Breviary (Figure 9-7). One of its miniatures depicts Saint Luke in the upper left corner seated at his easel painting the virgin and child. In the lower right—that is, at the other end of an imaginary diagonal—there is a large, magnificent dragonfly executed in typical Ghent-Bruges style.[19] It introduces a third object level: The first comprises miniature and text, the second, the borders with flower-filled strew patterns casting shadows on the golden background, and the third, the dragonfly that appears to have been attracted by the second-level objects and whose transparent wings extend both beyond the border zone into text and outward beyond the outer edge of the border. If Luke as the patron saint of painters is the emblem of artistic virtuosity, "the bravura depiction of the dragonfly . . . provides another kind of paradigm of the craft of painting."[20] One trompe-l'oeil is seduced by another. Even more remarkable is the doubling enacted in both corners of the miniature: The picture painted by Saint Luke in the upper left is an image-within-an-image that connects to the double trompe-l'oeil in the lower right. A layering of depth, a veritable visual *mise-en-abyme*, corresponds to the layering in the space in front of the initials and the border.

This dragonfly, or at least a copy of it, will reappear in Joris Hoefnagel. Even more so, it will put a finishing touch to the trajectory—the line of flight, as it were—of the trompe-l'oeil. First it appears as one of the marginalia added by Hoefnagel to the Book of Hours of Philipp of Cleves, then it resurfaces as an isolated, hyperrealistic insect specimen in Hoefnagel's *Animalia Rationalia et Insecta* (1575–80). The latter contains a further panel featuring realistically depicted dragonflies that truly cross the border into the real. As Thomas DaCosta Kaufmann and Virginia Roehrig Kaufmann have determined, the wings of the dragonflies are real; only the bodies are painted.[21] At the end of its trajectory the trompe-l'oeil has crossed the border separating the hallucinatory from the real. The thing itself materially emerges from the trompe-l'oeil.

FIGURE 9-7. Saint Luke painting the Virgin Mary and painted border with dragonfly. Attributed to the Master of the First Prayer Book of Maximilian. Page of the Grimani Breviary, ms. Lat. I 99 = 2138, folio 781v, Biblioteca Nazionale Marciana, Venice. Reproduced by permission of the Ministry of Heritage and Cultural Activities / Biblioteca Nazionale Marciana. Further reproduction prohibited.

The Imhof Prayer Book from 1511 contains trompe-l'oeils similar to those in the Grimani Breviary. Decorated by Simon Bening, a member of the inner circle of the Master of Mary of Burgundy, it was produced for the patrician Imhof family in Antwerp. Here we find a dragonfly that appears to be resting on a flower; its tail, however, extends beyond the frame, indicating that it is not sitting in but *on* the page. This is even more evident in the case of the fat black fly scurrying across the boundary between the border zone and the very edge of the page (Figure 9-8). The fly, so omnipresent in Dutch still lifes, has abandoned the space of writing and crossed over to our side, the space of the reader.

This self-conscious problematization of the coexistence of two-dimensional writing space and three-dimensional pictorial space also involved reflections on the verticality and horizontality of the surface of the page. In the wake of the restructuring of the manuscript page brought about by the Master of Mary of Burgundy and others from the Ghent circle, the verticality of the page is increasingly at odds with the objects that appear to be lying on it, which makes it necessary to double the image carrier by creating a fictitious object carrier alongside the real one.

The Legend of Saint Adrian (between 1477 and 1483) by the Master of the First Prayer Book of Maximilian contains a depiction of Louis XI and Charlotte of Savoy at the altar of Saint Adrian. It features a wreath of roses that appears to have been laid onto the ornamental pattern on the side of the page. Just as the page itself does not have to provide an intradiegetic explanation of why the letters stay in their place rather than drop down, it is equally unnecessary to explain why the wreath can lie on the page. The manuscript page is thus either a horizontal plane or a space devoid of gravity. Something else occurs in a page from the workshop of the Master of the Lübeck Bible depicting four female saints (Figure 9-9). Every single gem is hanging on a thread fastened to a wire cord serving as a kind of bearing frame. Something similar can be found in Marcus Cruyt's Book of Hours. Not only are the gems attached to the frame or to each other by a golden thread; some of them appear to have been pierced and fastened directly to the parchment by means of a golden wire. The manuscript page appears to have been interpreted as a vertical plane, that is, as a plane competing with the verticality of the panel painting.[22] Whatever resides on this plane is no longer *read* but *viewed*. If something is to stand or lie, it needs a horizontal surface. A part of the border must be transformed into a tabletop or something similar in order to maintain the intradiegetic continuum.

FIGURE 9-8. Page of the Imhof Prayer Book, 1511, by Simon Bening and the Master of the Scenes of David in the Grimani Breviary. Tempera on vellum, 9 × 6.2 cm. Private collection. Photo © Christie's Images / The Bridgeman Art Library.

FIGURE 9-9. Saint Catherine, Saint Clara, Saint Agnes, and Saint Barbara, from the workshop of the Master of the Lübeck Bible. Single folio, 23 × 16 cm. © Fitzwilliam Museum, Cambridge.

The Grimani Breviary offers one possible way of introducing such a resting surface. It consists of treating the colored manuscript page itself, which constitutes the background for the resting objects, as a flat picture object, the representation of a curvable parchment. In the lower left corner of folio 781v of the Grimani Breviary the golden border background turns into a flat picture object that recedes into the third dimension, thus revealing a horizontal plane on which a vase is standing. *Virgin and Child Crowned by an Angel* from a book of hours by the Master of the Dresden Prayer Book contains a similar example (Figure 9-10). The bottom of the painting resembles a drapery fold or an upright, slightly curved parchment, which effectively turns the material image carrier into a picture of itself. Drapery or parchment are pulled slightly backward to reveal a horizontal green ground supporting a flower vase and a snail. Here, the ambivalent space of the still life is starting to take shape. The snail clarifies matters: The border zone doubles itself; it is what it was and yet it is also its own representation or self-thematization that flips over into a horizontal position and thus turns first into a shelf, then into a niche, and finally into a table. The snail, which prior to the self-reflexive doubling of the ground and its move from vertical to horizontal would have been a trompe-l'oeil, can be deciphered as a "crypto-trompe-l'oeil."

The next step consists of equipping the entire border zone with an intradiegetic board structure that subsequently serves as a supporting plane for the objects distributed along its edge. Thus the border transforms into a niche or shelf and attains imaginary depth.

In the Hour Book of Engelbert of Nassau, the Vienna Master of Mary of Burgundy had already undertaken this step (Figure 9-11), which may serve as an example showing that the stages described here do not necessarily correspond to a chronological sequence. The border zone surrounding the miniature depicting the adoration of the Magi has been transformed into a niche cabinet. The niches or compartments contain a vase of peacock feathers, a decorated plate, saucers with berries or spices, a glass, and a porcelain carafe. In the case of Ambrosius Bosschaert the Elder, said to be one of earliest practitioners of the flower still-life genre, combinations of vases and bouquets are already classified as still life (Figure 9-12), though they are still linked to the niche, making evident the connection to the hour book genre. Still-life objects are linked to a niche architecture that establishes a connection between the picture object and the world of real things. And sure enough, around the edges and in wall openings we come across the little vanitas creatures: flies, spiders, beetles. They indicate that the niche edge once was a border that underwent the doubling

FIGURE 9-10. Master of the Dresden Prayer Book, *Virgin and Child Crowned by an Angel*. Staatliche Museen Preussischer Kulturbesitz, Berlin. © bpk / Bildagentur für Kunst, Kultur und Geschichte.

FIGURE 9-11. Vienna Master of Mary of Burgundy, *Adoration of the Magi*, c. 1475–80.
Page of the Book of Hours of Engelbert of Nassau, Ms. Douce 219-20, folio 145v,
Bodleian Library. By permission of The Bodleian Library, University of Oxford.

FIGURE 9-12. Ambrosius Bosschaert the Elder, *Flowers in a Glass Vase*, c. 1619. Oil on wood panel, 35.1 × 22.9 cm. Liechtenstein Museum, Vienna. © Liechtenstein. The Princely Collections, Vaduz-Vienna, inv. GE 57.

described above. The wall part of the niche is a metaphor for the parchment as the material image carrier, while the part of the niche that curves backwards is a metaphor for the receding surface of the parchment that (as in the case of the Grimani Breviary) has become its own image. The niche is one possible way of integrating ontologically heterogeneous elements into an apparently homogeneous picture object.

An Austrian book of hours from around 1500 shows a further step in the intradiegetic shift towards still life. The borders of a double page depicting the resurrection of Lazarus are interpreted as a shelf construction erected on top of a chest. Prefiguring the tabletops of later still lifes, the lid of the chest and the shelf compartments are filled with precious objects, gems, wreaths, and vanitas symbols including skulls, coins, a violet, and a peacock. The miniature itself is a window in the shelf allowing a view into the distance. On the opposing page the shelf's center window serves as frame for the text. As Otto Pächt noted, previously the written page contained the picture; now the picture contains the written page.[23]

A famous miniature in the Hours of Mary of Burgundy (attributed to Lieven van Lathem and Nicolaes Spierinc) offers a clear illustration of the way in which the border is transformed into a ledge designed to draw the observer into the devotional picture that constitutes the main part of the page (Figure 9-13).[24] The miniature's frame assumes the shape of a window that invites devotional contemplation. On the ledge there are several utensils associated with devotional practice that could very well be doubling the objects in the real space of the reader: a pillow to support genuflecting knees, a rosary, and of course the hour book itself, which by leaning against the picture object doubles the picture carrier, that is, the hour book that contains this miniature. The painted window ledge serves to unite into a homogeneous continuum the real space of the vision in front of and the envisioned space behind the window; similar to the space in Jan van Eck's *Madonna of Chancellor Rolin*, where the envisioned madonna and the chancellor appear to share one and the same reality (only initiates are able to decipher the madonna as a reality emanating from reading). Vision as the result of ecstatic reading generates the co-presence of visionary and vision. Where does the space of the reader end and the vision begin?

MEDIA COMPETITION

How are we to account for this refashioning of the manuscript page? How are we to explain this process by which the border is turned into a space connected to the real space of the reader and the miniature acquires an infinitely receding space of its own? Ultimately, this transformation arises from the

FIGURE 9-13. Lieven van Lathem and Nicolaes Spierinc, *Christ Being Nailed to the Cross*, c. 1475–80. Miniature on vellum, 22.5 × 16.3 cm. Page of the Book of Hours of Mary of Burgundy, codex 1857, folio 43v, Österreichische Nationalbibliothek. © Österreichische Nationalbibliothek, Vienna.

competition with reading practices enforced by the printed book. While Protestant reading practices in particular tended toward freeing the book from the presence of the reader's body, the illuminated manuscript appealed to the presence of the reader by attempting to have the medium itself dictate the practices linking book and reader, from the reading posture to the positioning of the page in space. Protestant cultural techniques of reading spiritualize the signified. "[F]aith is the energy that makes the signifier incandescent, until it dissolves its medial facticity."[25] By contrast, the hour books tend to create a space organized by thresholds that interlace zones of varying degrees of reality.[26] It is a space in which thresholds turn into relays, through and by means of which painted and written signs become the very things they signify—not as representations in the readers' mind but as hallucinations arising before their eyes. Reacting against the reading practices enforced by printed books, the trompe-l'oeils of the Ghent-Bruges style, the doubling of the real space of the reader wrought by the Master of Mary of Burgundy, not to mention Hoefnagel's excessive recto-verso trompe-l'oeils, invoke hallucinatory reading practices negating the boundary between the pictorial space and that of the reader's body. It is important to note, however, that hour books were engaged in developing strategies of hallucinating the referent even before the Protestant books set out to hypnotize the reader. The signifier is not vaporized in the glowing heat of faith; rather, in Benjamin's words, it appears as if the reality of co-presence had seared the subject.[27]

THE BIRTH OF REPRESENTATION

Dutch still life thus literally grew out of the edges of fifteenth-century hour books. It broke out of writing.

> [T]he birth of the new pictorial genres takes on the characteristic of a *distortion* or, if you like, a *cut*. Born as *marginalia*, a *reverse*, an *outside-the-work*, an *image-frame*, in a word as a *parergon*, still-life—in the seventeenth century—becomes an *ergon*.[28]

Still life is an undisciplined border zone that burst free and proceeded to conquer the entire space of representation.

With their strange mixture of transparency and opacity, Jan van Kessel's quodlibets and the model pages of the *Archetypa* are the result of a differentiation of the illuminated page: The border turns into a still life whose trompe-l'oeils are the not completely integrated parts of the illusionist Ghent-Bruges border fashioning. The ubiquitous tabletops, in turn, with their assorted objects and animals, emerged from the shelves and niches that—under pressure

from an intradiegetic disambiguation of the differing spaces created by proliferating writing ornaments extending into image space—from around 1480 on
had come to replace the old borders. The stone slab in Jan van Kessel's *Insects
on a Stone Slab* testifies to this origin. It depicts both the attempt to tip the border of the illuminated manuscript into the illusion of a three-dimensional central perspective *and* the resistance against this endeavor. Because it is impossible
to overcome the resistance of the two-dimensional page, van Kessel has great,
if not insurmountable difficulties trying to bring his quodlibets into a horizontal position.

What happens between the end of the fifteenth century and the ambivalent
dimensions of Joris Hoefnagel in the late sixteenth century and Jan van Kessel
in the seventeenth is nothing less than the dramatic, highly complex emergence
of the order of representation: "Representation depends on the disjuncture of
the discourse-figure couple."[29] Line, figure, and text of the illuminated manuscript belong to the order Jean-François Lyotard has called "the figural." Pictorial, ornamental, and textual space are intertwined; the figure is located on this
side of the signifier—the writing initially contained in the textual space can at
any point break out into ornamental proliferation, and the ornament of the margin, in turn, can be transformed into a representational picture.[30] From a media-
theoretical perspective it is necessary to add two observations. First, in the order
of the figural the sign carrier—the materiality and dimensionality of the
medium—cannot be separated from the sign itself. In the order of representation, however, sign and sign carrier, figure and ground, belong to different ontological categories and are therefore strictly separated. Metamorphoses are
therefore off limits. Second, the order of representation disallows any co-presence
of the body of the reader and the envisioned body engineered by trompe-l'oeil
borders. The abandoned co-presence is replaced by the classic sign model of
representation in which the represented is absent. This shift is based on the rejection of any hybridization of being, that is, of any intermingling of material
sign carrier and represented object. The medium must become invisible in order for the order of representation to override figural metamorphoses. What
pushed the figural towards representation was the increasingly problematic nature of the ground: Is it still a sign carrier or itself a picture object that obscures
the real ground of the signs? Not only are we dealing with a form of similarity,
in which, as described by Michel Foucault in *The Order of Things*, signs and perceptible objects are characterized by their similarity. This similarity itself is
caused by suppressing the difference between sign carrier (the edge of the parchment) and the sign or represented object. As the order of the figural disappears,
the order of representation is constituted. This occurs once the border ground
splits and doubles itself. It is now a figure of itself (a board) and a set of picture

objects that emerges from the ornamental proliferation of writing. The material carrier only reappears in this process of splitting and doubling in the shape of the trompe-l'oeil. The trompe-l'oeil—and it alone—bears witness to the order and practice of co-presence. In Aby Warburg's words, we may conceive of the trompe-l'oeil as an "afterlife" of the medium of the illuminated page. Representation, therefore, is not a semiotic issue; it has to be viewed as a process. It is a coding procedure, which, in turn, must be seen as a kind of excommunication (or suppression) of the formerly widespread hybridization of categories.

Van Kessel's hybridizations of bi- and tridimensionality as well as the metamorphic objects proliferating in the books of hours and prayer books by Hoefnagel and others are compromises between the finitude of the plane medium and the infinity of the imaginary three-dimensional perspectival space. In the trompe-l'oeils of the Ghent-Bruges manuscript illuminators, and especially in the proto–still lifes of Hoefnagel and the ambivalent spaces of Jan van Kessel, the age of representation is revealed to be an effect of a media epoch in which the self-affirming medium both endorses and negates its planarity by insisting on its two-dimensional structure while also offering a view of what lies beneath and beyond.[31]

To conclude, the trompe-l'oeil is anything but an additional ingredient that appears to be crossing the threshold from the illusionary pictorial to the real space. It indicates that the still life emerged from margins, edges, and borders that were constantly reinterpreted as represented objects. It is this oscillating between the transparency of the imaginary pictorial space and the opacity of the material carrier, and more importantly, it is the re-entry of the latter into the former, that keeps generating the trompe-l'oeil. The Dutch still life is the pictorialized, ongoing, unarrestable collapse of the distinction between material carrier and painted object. The trompe-l'oeil—the fly perched on the fruit, the plates, knives, and lemon peels sticking out over the edge of the table—is a symptom of the suppressed order of co-presence and the figural, and thus also a symptom of the media genealogy of the still life and representation itself.

DOOR LOGIC, OR, THE MATERIALITY OF THE SYMBOLIC

From Cultural Techniques to Cybernetic Machines

OPENING AND CLOSING DOORS

We have lost the ability, Theodor W. Adorno lamented in his American exile, "to close a door quietly and discreetly, yet firmly."[1] Adorno diagnosed the decline of this elementary cultural technique as nothing less than a prelude to fascism. One has to slam car doors and refrigerator doors, he observed, while other doors snap shut on their own. Doors cease to be cultural media that preserve a "core of experience" and instead change into machines that demand movements in which Adorno, in all seriousness, saw intimations of "the violent, hard-hitting, unresting jerkiness of Fascist maltreatment."

One can make of this what one will, but it does suggest that Adorno, insofar as he understood the disappearance of the door handle as an event of epochal significance, has to be counted among those philosophers of culture who already in the 1940s confronted the fundamental significance of cultural techniques. Adorno places gesture and mechanism, human and nonhuman actors into a relation in which both sides are invested with agency and in which the nonhuman actor has the power to decenter and disable the very being of the human subject. This concept of cultural techniques, however, implies a definition of culture as "refinement." For Adorno, culture is something that only pertains to people who associate with things anthropomorphically. Cultural techniques, then, would be gestures that anthropomorphize things, that include them in the humanoid sphere as products of the nonalienated work of craftsmanship.

The notion of cultural techniques proposed in this volume is based on a different concept of culture. It implies a plurality of cultures and abandons all one-sided conceptions of human-thing relations that privilege humans. Contrary to Adorno's anthropocentric assumption, culture is a humanoid-technoid hybrid. It has always been so (and not only since the invention of the automatic door). Cultural techniques inevitably comprise a more or less complex actor-network that includes technical objects and chains of operations (including gestures) in equal measure. The "human touch," the power of agency typically ascribed to humans, is not a given but is constituted by and dependent on cultural techniques. In this sense, cultural techniques allow the actors involved to be both human and nonhuman; they reveal the extent to which the human actor has always already been decentered *by* the technical object. In other words (those of Lacan), cultural techniques point to a world of the symbolic, which is the world of machines.[2] The door is—or was—such a machine. Discussing doors—opening, as it were, a passage into a new understanding of doors—may demonstrate how cultural techniques consist of chains of operations including body techniques, which makes it possible to conceive of image spaces as constituted by virtual media operations.

FORES AND NOMOS

As mentioned in the introductory chapter, every culture starts with the introduction of distinctions. Speaking in systems-theoretical terms, this presupposes both an observer who observes this distinction and a set of techniques that process the distinction and thereby render observable the unity of the things distinguished. Thus the difference between humans and animals is one that depends on the mediation of a cultural technique. In this not only tools and weapons (which paleoanthropologists like to interpret as the exteriorization of human organs and gestures) play an essential role, but also the invention of the door, whose first form was presumably the gate (*Gatter*)[3]—which is hardly an exteriorization of the human body. The door appears much more as a medium of the coevolutionary domestication of animals *and* humans. The construction of a fold with a gate, something that turns the hunter into a shepherd, leads not only to the domestication of animal species but above all to the interruption of those human-animal metamorphoses that Paleolithic cave paintings attest to.[4] Already Gottfried Semper recognized the fold as "the most original vertical spatial enclosure [*Abschluss*] invented by man."[5]

Doors and door sills are not only formal attributes of Western architecture, they are also architectural media that function as cultural techniques because they operate the primordial difference of architecture—that between inside and outside.[6] At the same time they reflect this difference and thereby establish a system comprised of opening and closing operations. In contrast to the "mute" closedness of the nonarticulated wall, Georg Simmel wrote, the closed door is both closed and the sign of this closedness.[7] The door emphasizes the unity of the difference between inside and outside, since "it shapes the possibility of closure against the backdrop of the possibility of opening and keeps virtually present both possibilities."[8]

With regard to the town gate and its relation to the law, the primordial function of the door—in the sense of Latin *fores* or Greek *thýra*—could be labelled nomological. The Greek *nomos*, usually translated as "law," is connected to the concrete operation of land division. According to Carl Schmitt, it separates a circumscribed space from an outside, thus creating a difference on the basis of which political, social, and religious orders can be established. But as Kafka's famous parable "Before the Law" makes clear, the law is constituted in the first place by an opening that grants access to the law. A door is a place where the difference that constitutes the law has to negate itself in order to become effective. A man from the country waits in front of an open door, the first of a multitude of doors, which is closed by the symbolic order enacted by the gatekeeper.[9] In Kafka's text the usually alternating states of being either closed or open overlap each other. Is the door closed while it is open? The waiting of the man in front of the door creates the paradox that a state of being open causes the effect of an interruption.[10] The logic of a door that is closed while it is open is the logic of the symbolic. The door and the gatekeeper implement the differential law of the signifier itself. To step through a door, in turn, is to submit to the law of a symbolic order, a law—be it the law of the *polis* or the paternal *oikos*—that is established by means of the distinction between inside and outside. A door, Lacan insisted, is nothing altogether real; on the contrary. "In its nature, the door belongs to the symbolic order. . . . The door is a real symbol, the symbol par excellence, the symbol in which man's passing, through the cross it sketches, intersecting access and closure, can always be recognized."[11]

"Fores . . . in liminibus profanarum aedium ianuae nominantur," Cicero noted: "Doors are called the access points [*ianuae*] at the thresholds of profane buildings."[12] The door is intimately connected to the notion of the threshold, a zone that belongs neither to the inside nor the outside and is thus an extremely dangerous place. The Romans saw the house door as dividing two worlds, "the world outside, where there are innumerable hostile influences and powers, and the region within the limits of the house, the influences and powers of which are friendly."[13] Arnold van Gennep interprets crossing through doors and gates as a rite of passage: "To cross the threshold is to unite oneself with a new world. It is thus an important act in marriage, adoption, ordination, and funeral ceremonies."[14] Many sarcophagi and funeral altars depict house doors or city gates. Just as every bridge points to that last bridge leading into the beyond, so every threshold points to that last threshold at the entrance to Hades, which mortals at the end of their earthly sojourn must cross over, whether to the gates of hell below or the pearly gates above.

Since very early on, then, the processing of the distinction between inside and outside has been tied to ways of operating the distinction between sacred and profane zones, and this may well be the first of all cultural articulations of space. It is not surprising that the door sill is of a sacred nature. It is framed by countless security measures: horseshoes, images of Saint Sebastian, the souls of animals sacrificed on the sill, the corpses of enemies buried beneath it, a particular roof top, a container for holy water, the Mezuzah, or even plain doormats.[15] The door sill is eerie. No wonder the ethnologist Marcel Griaule once described the door as a "terrible instrument," which one "must not use except according to certain rites and with a pure conscience, and which one has to invest with all kinds of magical guarantees."[16]

THE DOOR LOGIC OF THE *MÉRODE TRIPTYCH*

Early fifteenth-century cultural artifacts from the Flemish nobility and rich bourgeoisie appear to be obsessed with enacting the difference between sacred and profane zones by employing interconnected folding operations that are differentiated into a multitude of cultural techniques. The *Mérode Triptych* by

FIGURE 10-1. Robert Campin, *Mérode Triptych*, 1425–28. Oil on wood. The Cloisters, Metropolitan Museum of Art, New York. © bpk / The Metropolitan Museum of Art.

Robert Campin, who was known also as the Master of Flémalle (Figure 10-1), demonstrates how the door initiates a sequence of operations that connect the working of the difference between opening and closing with working both the difference between the sacred and the profane and spiritual vision and profane appearance. The central panel recalls the *Annunciation* by the same Master in the Musées Royaux des Beaux-Arts de Belgique in Brussels (Figure 10-2). Closer scrutiny reveals that the *Mérode Triptych* includes as *real object* what in the *Annunciation* had still been a *picture*. This real object works as a foldable object and thus as a cultural technique.

Let us take a closer look. The Virigin Mary is sitting in front of a wooden bench with a foldable back. The chimney wall features a colored woodcut showing Saint Christopher and two candle holders, one of which is folded towards the wall. The table can be folded as well. Beyond the open windows we do not see a landscape but a gold ground, consisting not of gold leaf but of a metal foil covered with a yellow varnish. Mary's robe bears an inscription representing the beginning of the angelic salutation, "Salve Regina." The Master of Flémalle was fully aware of the close relationship between books and textiles: A book could easily be transformed into a garment by the mediation of a fold. Moreover, precious books of that time were often girdle books, much like the one lying on the table. Because books often possessed their own "gown," books and gowns or robes belonged together. Behind the archangel Gabriel there is an open door. One can clearly discern the floor of the corridor outside, and it was through this opening that the angel entered. Mary is reading a book of hours; another one, already mentioned, lies on the table, opened in such a way that it shows the different ways in which the pages are folded.

The central panel of the *Mérode Triptych* (Figure 10-3) also depicts the annunciation, but it includes a number of significant variations. First, instead of the gold ground in the window, the observer sees part of the sky. Second, in lieu of the second window we have a niche with gothic tracery and a kettle. Third and most significant, the door behind the angel has been replaced by two circular windows. So how did Gabriel enter? The door of the Brussels *Annunciation* is not just deleted, it has been intradiegetically and categorically displaced. There is also a door in the left wall, but instead of opening into another painted room as in the painting in Brussels, it establishes a connection to the pair of donors who are depicted on the left panel attending the annunciation.[17] The hinges connecting the left panel to the centerpiece are precisely where the painted door panel would be attached to its jamb, of which we can see a part on the very left of the central panel. Thus the painted door signals to the observer the meaning of the real panel on which it, the door, has been painted. The panel signifies an opening that allows the observer to apprehend a higher, numinous reality. The

FIGURE 10-2. Robert Campin, *Annunciation*, c. 1420. Tempera on oak, 63.7 × 61 cm.
© Musées Royaux des Beaux-Arts de Belgique, Brussels. Photo: J. Geleyns / Ro scan.

FIGURE 10-3. Robert Campin, *Mérode Triptych*, central and left panels. © bpk / The Metropolitan Museum of Art.

doubling of the folding apparatus thereby initiates a game of presence and absence: The unfolding of the left panel of the triptych grants the donors access to the sacred proceedings which are both absent and present. In classical terms of representation the annunciation is absent, given that we are dealing with the pictorial representation of a past event; as a *vision*, however, which—to pick up on Benjamin's famous phrase—has entered the age of its mechanical reproducibility, it is present. The intertwining of the real and the fictitious foldable object at the point where real and painted hinges meet leads to a merging of presence and absence. Taking part in this game is also the open door in the outer wall that encloses the garden and is guarded by a messenger of the city of Mecheln. Behind it one sees the houses across the street with their doors and windows. On the right panel we see the foster-father of Jesus, Saint Joseph, drilling holes into a piece of wood. On the table there is a recently completed mousetrap; another one sits on the shutter that has been folded outwards and serves as a sales counter for the display of merchandise. Other shutters have been folded upwards under the ceiling. It is a space organized by hinges, with all kinds of folding operations that open up spaces in every possible directions: in, out, and beyond.

In highly conspicuous and tangible ways, the very process of visual perception is here connected with the opening and closing of various media: turning pages and panels, opening doors and books, "unfolding" the triptych itself. These operations are especially interesting since they directly concern the processing of the symbolic and the imaginary, but it is no less important to understand how these operations are taken up and escalated by further folding operations: the foldable back of the bench, the shutters, the candle holders, the iconographically notorious mousetrap. Most of these foldable and collapsible objects are staged in such a way that their foldability is displayed in conspicuous fashion. These visual demonstrations—in the first instance, of course, the operative chain linking door, book, and triptych—reveal the extent to which reading is thought of and represented as equivalent to seeing, and seeing, in turn, as equivalent to revealing.[18] Media come to negotiate the slippery boundary between gaze and vision: What is at stake here is a seeing that is also a revealing—a revealing of something that is not just simply given and therefore has to be made to appear by the process of folding.[19] Revelation itself becomes a sequence of architecturally based medial transformations. It is as if the house as a whole had been seized by this way of seeing.

If one looks at the complex "devotional apparatus"[20] on display in fifteenth-century transalpine culture from the perspective of a theory of the door as a cultural technique, unfolding appears to be a highly differentiated technology designed to simultaneously articulate and operationalize the precarious threshold between appearance and vision, the profane and the sacred. "Here the possibility of bridging the gap between the profane and the sacred is suggested insofar as by the practice of folding, *seeing* is staged as *appearing*."[21] Karl Schade had something very similar in mind when he asked whether early Dutch triptychs should be understood as a kind of staging apparatus, the object form of which was able to produce "almost-visions."[22] That would imply first of all that the desire of the spectator to exchange glances with the numinous is transposed in a fetishist manner onto the level of the medium and its way of operations.[23] Representation in the *Mérode Triptych* appears as an actualization of virtual operations of folding. If Italian Renaissance painting since Alberti is defined as a section through the pyramid of vision, then one can possibly define early Dutch painting as a section through an accumulating series of hypothetical unfolding operations.

Here, then, is the follow-up thesis: In an attempt to recode seeing as appearing into seeing as voyeurism, Dutch interior painting of the seventeenth century took up and redirected the mediality of the folding apparatuses of Flemish art, which served to recast the numinous as the intimate. On Jan Steen's *The Morning Toilet* of 1663 (Figure 10-4), the real surface of the image is interpreted by a diaphragm in the shape of a grandiose portal of columns. Following their invention by the Master of Boucicaut, these diaphragms had often been used by Rogier van der Weyden, for instance, and for framing miniatures in books of hours. Thus Steen is consciously taking up the tradition of bridging the gap between seeing and vision. But while such diaphragms in the fifteenth century did not allow for the operations of opening and closing, simply because those portals were lacking door panels, Steen attached to the inner side of the portal a very common door that (together with a pan of the whole situation by 90 degrees, putting us into the place of the donors) allows us to look into a bedroom and watch a woman get dressed. The voyeuristic character of this gaze through the open door is connected in a recursive way to the opened bed curtains, which add to the threshold of the image yet another one that interprets and comments on the former. Steen's strange opening works like an interface between the ages. The gaze that had been shaped by religious visual and textual media in the fifteenth and early sixteenth century is now molded by means of a special double-sided door into a form of seeing highly characteristic of Dutch seventeenth-century interiors: It is transformed by the interconnection of the front and the back sides of the diaphragm that define how one has to interpret the surface of the image. The Dutch painting is constituted by flights of thresholds, commenting on each other.[24]

FIGURE 10-4. Jan Steen, *The Morning Toilet*, 1663. Oil on wood, 53 × 64.7 cm. Windsor Castle, London. Royal Collection Trust / © Her Majesty Queen Elizabeth II, 2013.

POST–DOOR-LOGICS AND *THEIR* VISIONS

"Doors," Robert Musil wrote in 1926, "are a thing of the past." The door's moveable board, set in the wall, "has already lost most of its significance. Up until the middle of the last century you could listen in with your ear pressed against it, and what secrets you could sometimes hear!"[25] What Musil had in mind was a door-logic that still corresponded to that depicted in Dutch interior paintings by Steen or van Hoogstraten. According to Musil the door had been the site of dramas—it enabled eavesdroppers to become privy to dramas of disinheritance, secret marriages, and devious plans to poison the hero. But these sites of initiation are no more. As long as doors functioned as operators of difference between inside and outside, they also helped to create, in line with the public-private distinction, an asymmetry of knowledge. Doors produce an information gap; they play an indispensable role in the production of thermodynamic or information-theoretical knowledge. Not by chance is Maxwell's demon a gatekeeper.[26] As long as doors fulfill their informative function, they sustain a disequilibrium of energy or knowledge that defers overall entropy. In this way doors are crucial actors in the distribution and circulation of knowledge. In modern concrete buildings, however, doors have surrendered that function to walls. Walls have turned into membranes, so that one can only wonder, with Musil, "Why has no radio-poet yet taken advantage of the possibilities of the modern concrete structure? It is undoubtedly the predestined stage for the radio play!"[27] If walls, as Musil surmised, have become membranes in modern living-machines, then doors lose the function Simmel ascribed to them: to signify the closedness of the wall on the basis of their virtual opening. The information differential is balanced out. Maxwell's demon is wrecked; entropy reigns. In a situation of complete entropy nothing more can happen; whereas that more can happen could yet be asserted of the classic form of narrative.

"The only original door conceived by our time," Musil writes, "is the glass revolving door of the hotel and department store."[28] The revolving door was invented in 1888 by Theophilus Van Kannel, whose patent application refers to it as the "new revolving storm door."[29] With the installation of this innovative contraption consisting of three or four panes of glass inside a circular wind-trap cylinder, it becomes possible to say that one is "in the door." In the past, Musil writes, entry doors had representational duties. The nomological door enacted a symbolic order to which one was subjected by crossing the threshold; the revolving door, on the other hand, is a biopolitical device for managing humans in motion. It imposes uniform speed on flows of people while separating those who enter from those on the outside. "In the old way," Van Kannel boasted of his invention, "every person passing through first brings a chilling gust of wind with its snow, rain, or dust, including the noise of the street; then comes the unwelcome bang."[30] The revolving door represents a reinterpretation of architecture as a thermodynamic and hygienic machine with an attendant change from nomological to control functions. Basically it constitutes a paradox: One walks through a door that is permanently closed. "Always Closed" was in fact Van Kannel's first advertising slogan (Figure 10-5). Finally, there is one very obvious feature that links the revolving door to the

FIGURE 10-5. "Always Closed." Reprinted from Robert Blanchard, *Around the World with Van Kannel* (New York: Van Kannel Revolving Door Company, 1930).

disappearance of the door from human life: the absence of the door handle. Neither revolving nor sliding doors have handles or knobs. Maybe it is possible to define the epoch of bourgeois architecture as the epoch of the door handle? By virtue of the latch the door is a tool that demands to be operated by the hand of a user.

Now that such transit zones as luxury hotels, department stores, cars, and ocean liners have replaced the house as sites of social representation, "there is no house" anymore. [31] Adorno stated almost the same in 1944 with regard to the end of the transcendental possibility to dwell: "Actually man can no longer dwell."[32] Our social status is no longer represented by our way of dwelling but by the places where we stop off for the night and the vehicles we inhabit. Adorno realized the vanishing of the possibility to dwell when he learned about American trailer parks and mobile home estates: "The hardest hit, as everywhere, are those who have no choice. They live, if not in slums, in bungalows that tomorrow may already be leaf-huts, trailers, cars, camps, or the open air. The house is past."[33]

In 1929 Le Corbusier installed sliding doors in his Maison Loucheur, thus turning the living room into several sleeping compartments. The sliding door was "the lever which started the *machine à habiter*,"[34] and finally relegated the house to the past. Not by chance did space-saving sliding doors first appear as cabin doors on ocean liners and in railway compartments. Sliding doors are the signature of an era in which architecture is subject to the dictates of transit rather than the rules of dwelling. Fully automated sliding doors were first attempted in 1896. From 1914 on they were equipped with motors or hydraulic systems. In the 1930s the automation of the door was perfected with optical sensors such as the "electric eye" or pressure-sensitive "magic carpets." As a result the "responsibility for the opening and the closing of a door had been completely transferred from human to machine."[35] Operations of opening and closing, once clearly within the domain of human agency, bid farewell to man. Only by an act of mercy do sliding doors open in front of approaching human actors whom they have downgraded to mere agents of their opening. No longer do they receive their orders from the person desiring to pass through; now an invisible power rules over their opening and closing.

As mentioned above, the logic of a door which—as in Kafka's parable of the gatekeeper—is closed while it is open is the logic of the symbolic. The door and the doorkeeper incarnate the differential law of the signifier itself. The twentieth century has invented doors that implement this logic by literary as well as by electronic means. Compared to their replacements, traditional doors truly are things of the past. Modern doors have irretrievably forfeited their nomological for a cybernetic function. The basic distinction of inside and outside has been replaced by the distinction between current/no current, on/off. The cybernetic logic of opening and closure estranges the old nomological logic: The electronic door, the switching element, is a door "where," to put it in Lacan's words, "something passes when it is closed, and doesn't when it is open."[36] Lacan added that what is important with regard to cybernetic doors "is the relation as such, of access and closure. Once the door is open, it closes. When it is closed, it opens."[37] The technical name of the logical switching circuits that were made of these doors is *gates* (*Gatter* in German), a name that recalls the ancient cultural-technical meaning of doors. But these gates do not open into an outside or the animal domain. They open themselves by being closed only to other gates and/ or to themselves. Reality appears to have become more and more psychotic.

From Freud to Lacan, psychoanalysis has taught us that the distinction between inside and outside is at the basis of the constitution of reality. In other words, any reality check that culminates in existential judgments such as "this object is real, it exists in reality" always operates against the background of the complementary negative judgment, "this is not a dream, I am not hallucinating this."[38] But if the symbolic order is repudiated (as in the case of psychosis), reality takes on hallucinatory features. The imaginary then appears within the real or gets mixed up with it. In David Cronenberg's film *Videodrome* (1983), a paradoxical search for the origin of certain TV images said to be real rather than mere TV images leads to an elusive father figure, whose very name—Brian O'Blivion—indicates the degree to which the *nom-du-pere*, the name and the law of the father, which denies access to the forbidden object of desire, has been deleted from the symbolic order. "Your reality is already half video hallucination," says O'Blivion, who exists only as a video image. In *Videodrome* the concept of "vision," so crucial for the devotional apparatuses of old Netherlandish art, is taken up in both its religious and technical sense. In the film O'Blivion explains:

> The television screen is part of the physical structure of the brain. Whatever appears on the TV screen emerges as raw experience for those who watch it. Therefore TV is reality and reality is less than TV. Your reality is already half video hallucination . . . I had a brain tumor and I had visions. I believe the visions caused the tumor and not the reverse. I could feel visions coalesce and become flesh, uncontrollable flesh. But when they removed the tumor, it was called videodrome.[39]

Videodrome can be read as a reading of the *Mérode Altarpiece* or other fifteenth-century vision machines. Once a cultural technique that processed the difference between inside and outside and thus the transition from our world to the world of the numinous, the doors of the electronic age have turned into cybernetic machines, in the course of which the numinous is replaced by psychosis and vision by hallucination.

The film's main protagonist, Max Renn (James Woods), starts to hallucinate after having been subjected to the broadcast of extremely violent images. These hallucinations, however, are not clearly identified as such for the film's viewers by employing the usual cinematic conventions. On the contrary. Both the audience and Max lack means and criteria to distinguish between hallucination and reality. Inside and outside are inextricably short-circuited for Max once he is turned into his own video recorder. This is a very special video recorder indeed: Not only does it produce the images it plays, it watches them too. As a result, the audience watching *Videodrome* is watching what Max is watching; it enters the scene of the unconscious. The structure of the film is first person singular; this means that the I, which in "normal" reality acts as a shifter constituted by the symbolic order, has lost its ability to shift: "Only I am an I," as Roman Jakobson once put it with regard to an example of a psychotic breakdown of the shifter function.[40] Lacan conceded a crucial function to the concept of the shifter in explaining hallucinations. The narration from the perspective of the first person singular corresponds to an "I" that cannot turn

FIGURE 10-6. Still from David Cronenberg's *Videodrome* (Canada, 1983).

into a "You." Therefore, what is perceived appears strictly as reality. In the end, the daughter of the elusive father figure, Bianca O'Blivion, tells Max: "You have become the video word made flesh." Max has now run full circle through a closed Moebius loop, the two sides of which are reality and hallucination, the real and the imaginary, which constantly and imperceptibly merge into each other. He has now himself become the hallucinatory real. The cybernetic doors of the twentieth century create a psychotic reality that interprets itself in terms of the devotional folding apparatuses of the fifteenth century.

Not by chance is there a short sequence in *Videodrome* in which doors that have been lifted off their hinges are carried across a street by two workers (see Figure 10-6). The scene illustrates the characteristic feature of the age of electronic media: Signifiers are processed by doors that have completely "unhinged" the nomological function of doors. The Moebius loop structure of the film, which makes it impossible to decide whether we are inside "reality" or inside a hallucination, corresponds to the cybernetic feedback loops of pairs of electronic doors, or flip-flops, in which one door triggers the opening of another by its own closing, and vice versa. In a space where inside and outside are thus folded, wired, and coupled into each other as feedback loops, doors as cultural techniques have lost their moorings. Doors indeed belong to the past. Cronenberg's *Videodrome* teaches us how difficult/impossible it has become to determine whether the perception of a thing corresponds to an inner or outer reality. With the retreat of the symbolic from the constitution of reality, and with the difference between inside and outside losing its form, the place of the law is replaced by a short circuit between the imaginary and the real. Lacan expressed where this is leading to: No one knows anymore whether a door opens to the imaginary or to the real.[41] We are all unhinged.

NOTES

TRANSLATOR'S NOTE

1. For recent introductions to the concept, see Geoffrey Winthrop-Young, "Cultural Techniques: Preliminary Remarks," *Theory, Culture and Society* 30, no. 6 (2013): 3–19, and Bernard Dionysius Geoghegan, "After Kittler: On the Cultural Techniques of Recent German Media Theory," *Theory, Culture and Society* 30, no. 6 (2013): 66–82.

INTRODUCTION: CULTURAL TECHNIQUES, OR, THE END OF THE INTELLECTUAL POSTWAR IN GERMAN MEDIA THEORY

1. Ernst Cassirer, *The Philosophy of Symbolic Forms.* Volume 1: *Language*, trans. Ralph Manheim (New Haven: Yale University Press, 1955), 80 (translation modified).

2. On the discussion of a "German" media theory, see Eva Horn, "'There Are No Media,'" *Grey Room* 29 (2007): 6–13; Geert Lovink, "Whereabouts of German Media Theory," in *Zero Comment: Blogging and Critical Internet Culture* (New York and London: Routledge, 2008), 83–98; and John Peters, "Strange Sympathies: Horizons of German and American Media Theory," in Frank Kelleter and Daniel Stein, eds., *American Studies as Media Studies* (Heidelberg: Winter, 2008), 3–23.Further to the connection between media theory and cultural techniques, see Geoffrey Winthrop-Young, "Cultural Techniques: Preliminary Remarks," *Theory, Culture and Society* 30, no 6 (2013): 3–19.

3. Participants included Norbert Bolz, Wolfgang Coy, Charles Crivell, Wolfgang Hagen, Jochen Hörisch, Friedrich Kittler, Joachim Paech, Georg-Christoph Tholen, and myself.

4. See Hans Ulrich Gumbrecht, "Flache Diskurse," in *Materialität der Kommunikation*, ed. Hans Ulrich Gumbrecht and K. Ludwig Pfeiffer (Frankfurt/M: Suhrkamp, 1988), 919; see also Gumbrecht, "A Farewell to Interpretation," in *Materialities of Communication*, ed. Hans Ulrich Gumbrecht and K. Ludwig Pfeiffer (Stanford: Stanford University Press, 1994), 399.

5. See Jacques Derrida, *Of Grammatology*, trans. Gayatri Chakravorty Spivak (Baltimore: Johns Hopkins University Press, 1976), part 2, chapters 2, 3, and 4.

6. See Friedrich A. Kittler, *Discourse Networks, 1800/1900*, trans. Michael Metteer and Chris Cullens (Stanford: Stanford University Press, 1990), 27–53.

7. See Jacques Derrida, *The Post Card: From Socrates to Freud and Beyond*, trans. Alan Bass (Chicago: University of Chicago Press, 1987). Further see Geoffrey Winthrop-Young, "Going Postal to Deliver Subjects: Remarks on a German Postal Apriori," *Angelaki* 7, no. 3 (2002): 143–58.

8. From Paul Virilio, *Bunker Archeology*, trans. George Collins (New York: Princeton Architectural Press, 1994) to Paul Virilio, *War and Cinema: The Logistics of Perception*, trans. Patrick Camiller (London: Verso, 1989).

9. See, among others, Bruno Latour and Steve Woolgar, *Laboratory Life: The Construction of Scientific Facts* (Princeton: Princeton University Press, 1986); Bruno Latour, "The 'Pédofil' of Boa Vista: A Photo-Philosophical Montage," *Common Knowledge* 4, no. 1 (1995): 144–87; Hans-Jörg Rheinberger, *Toward a History of Epistemic Things: Synthesizing Proteins in the Test Tube* (Stanford: Stanford University Press, 1997); and Henning Schmidgen, *Hirn und Zeit: Geschichte eines Experiments 1800–1950* (Berlin: Matthes und Seitz, 2014).

10. For epochē see Heidegger, "Time and Being," in *On Time and Being*, trans. Joan Stambaugh (Chicago: University of Chicago Press), 1–25; for the epochē of media see Derrida, *Post Card*.

11. See Karl Knies, *Der Telegraph als Verkehrsmittel: Mit Erörterungen über den Nachrichtenverkehr überhaupt* (Tübingen: Laupp, 1857); on the concept of secondary orality, see Walter J. Ong, *Orality and Literacy: The Technologizing of the Word* (London and New York: Methuen, 1982).

12. See Cornelia Vismann, "Cultural Techniques and Sovereignty," *Theory, Culture and Society* 30, no. 6 (2013): 83–93.

13. See Joseph Vogl, "Becoming-media: Galileo's Telescope," *Grey Room* 29 (2007): 14–25.

14. See Bernhard Siegert, *Relays: Literature as an Epoch of the Postal System*, trans. Kevin Repp (Stanford: Stanford University Press, 1999).

15. See N. Kathrine Hayles, *How We Became Posthuman: Virtual Bodies in Cybernetics, Literature, and Informatics* (Chicago: University of Chicago Press, 1999), 251, and her "Cybernetics," in *Critical Terms for Media Studies*, ed. William J. T. Mitchell and Mark B. N. Hansen (Chicago: University of Chicago Press, 2010), 145–56.

16. This account is, of course, (all too) simplistic. Germany, too, witnessed a broad reception of postcybernetic American theories of "transhumanism," e.g., Donna Haraway's *Cyborg Manifesto*. However, this reception occurred especially where the critique of a McLuhanese media anthropology of extensions and prostheses was lacking, as in art or gender theory.

17. See the more elaborate characterization by Geoffrey Winthrop-Young, "Krautrock, Heidegger, Bogeyman: Kittler in the Anglosphere," *Thesis Eleven* 107, no.1 (2011): 6–20. Winthrop-Young refers to David Wellbery's criticism of Derrida's reception in the United States, as sharp as it is brilliantly formulated, in Wellbery's foreword to Kittler, *Discourse Networks, 1800–1900*, viii.

18. I do not know, however, whether this justifies our speaking of an "anthropological turn" in German media theory. See Erhard Schüttpelz, "Die medienanthropologische Kehre der Kulturtechniken," *Archiv für Mediengeschichte* 6 (2006): 87–110.

19. E.g., see the introduction to Cary Wolfe, *What Is Posthumanism?* (Minneapolis: University of Minnesota Press, 2010).

20. See Geoffrey Winthrop-Young, "Mensch, Medien, Körper, Kehre: Zum posthumanistischen Immerschon," *Philosophische Rundschau* 56, no. 1 (2009): 1–16.

21. On the media function of animals, see, e.g., Manfred Schneider's analysis of the secretarial function of dogs, "Das Notariat der Hunde: Eine literaturwissenschaftliche Kynologie," *Zeitschrift für deutsche Philologie* 126 (2007): 4–27. On animal sacrifice, see Thomas Macho, "Tieropfer: Zur Geschichte der rituellen Tötung von Tieren," in *Mensch und Tier: Eine paradoxe Beziehung*, ed. Stiftung Deutsches Hygiene-Museum (Ostfildern-Ruit: Hatje Cantz, 2002), 54–69. On the *fides* of dogs and *imitatio* of apes, see Gerhard Neumann, "Der Blick des Anderen: Zum Motiv des Hundes und des Affen in der Literatur," *Jahrbuch der deutschen Schillergesellschaft* 40 (1996): 87–122.

22. Cary Wolfe's analysis of "animal *Dasein*" arguably comes closest to the perspective adopted by cultural techniques research. See Wolfe, *Before the Law: Humans and Other Animals in a Biopolitical Frame* (Chicago: University of Chicago Press, 2013).

23. See Hartmut Böhme, Peter Matussek, and Lothar Müller, *Orientierung Kulturwissenschaft: Was sie kann, was sie will* (Reinbek: Rowohlt, 2000), 165.

24. A hundred years ago *Kulturtechnik* as an academic discipline would have been housed in agricultural or geoscientific institutes. As defined by the sixth edition of *Meyers Grosses Konversationslexikon* (1904), cultural techniques comprise "all agricultural technical procedure informed by the engineering sciences that serve to improve soil conditions," such as irrigation, drainage, enclosure, and river regulation.

25. One could also speak of empirical transcendentals.

26. In the wake of Levinas, David Wills writes that "[t]he house is therefore required to be *in* the world of objects without being *of* it; objects, including buildings themselves, are produced 'out of . . . a dwelling.'" David Wills, *Dorsality: Thinking Back through Technology and Politics* (Minneapolis: University of Minnesota Press, 2008), 56. Following Heidegger, the house would be a "thing" rather than an "object."

27. As a result, in the nineteenth century music was designated as part of culture only if it could be properly notated (and thus subjected to the alphabetical code). Painting, in turn, only acquired cultural status if it remained accessible to iconographic interpretation (which, in turn, pointed to the domain of books).

28. See Schüttpelz, "Die medienanthropologische Kehre der Kulturtechniken," 90.

29. Thomas Macho, "Zeit und Zahl: Kalender- und Zeitrechnung als Kulturtechniken," in *Bild—Schrift—Zahl*, ed. Sybille Krämer and Horst Bredekamp (Munich: Wilhelm Fink Verlag, 2003), 179.

30. Gumbrecht, *A Farewell to Interpretation*, 402.

31. The phrase *cultural technique* thus connects with Bruno Latour's media-theoretically informed notion of *immutable mobile*. Replacing the metaphysical confrontation of language and world, both cultural techniques and immutable mobiles

introduce discontinuous series of operations that transform things into signs, a process vital for the ways in which knowledge functions and produces evidence.

32. Vismann, "Cultural Techniques and Sovereignty," 83.

33. Thomas Macho, "Second-Order Animals: Cultural Techniques of Identity and Identification," *Theory, Culture and Society* 30, no. 6 (2013): 31.

34. Macho, "Second-Order Animals," 31.

35. See Claude Lévi-Strauss, *The Origin of Table Manners*, vol. 3 of *Introduction to a Science of Mythology*, trans. John and Doreen Weightmann (New York: Harper and Row, 1978), 478–90.

36. Thomas Macho, *Vorbilder* (Munich: Wilhelm Fink Verlag, 2011), 45.

37. See Tim Ingold, "Toward an Ecology of Materials," *Annual Review of Anthropology*, 41 (2012): 427–42, especially 438.

38. See Jacques Derrida, "Signature Event Context," trans. Samuel Weber and Jeffrey Mehlman, in *Limited Inc* (Evanston, Ill.: Northwestern University Press, 1977), 1–23.

39. Gottfried Semper, *Style in the Technical and Tectonic Arts; or, Practical Aesthetics*, trans. Harry Francis Mallgrave and Michael Robinson (Los Angeles: Getty Research Institute, 2004), 242. It is important to note that the German word *Wand* (wall) is etymologically closely related to *Gewand* (garment). See also Bernhard Siegert, "After the Wall: Interferences among Grids and Veils," *Graz Architektur Magazin* 9 (2012), 18–33.

40. See Marcel Mauss, "Techniques of the Body," in *Incorporations*, ed. Jonathan Carey and Sanford Kwinter (New York: Zone Books, 1992), 454–77.

41. See Harun Maye, "Was ist eine Kulturtechnik?" *Zeitschrift für Medien- und Kulturforschung* 1 (2010): 135.

42. See Michel Serres, *The Parasite*, trans. Lawrence R. Schehr (Baltimore: Johns Hopkins University Press, 1982), 53.

43. Lévi-Strauss, *Origin of Table Manners*, 489. On eating rituals as a cultural technique, see chapter 2 in this volume: "Eating Animals—Eating God—Eating Man," p. 33.

44. See Ute Holl, "Postkoloniale Resonanzen," *Archiv für Mediengeschichte* 11 (2011): 115–28.

1. CACAPHONY OR COMMUNICATION? CULTURAL TECHNIQUES OF SIGN-SIGNAL DISTINCTION

1. For an overview see Oswald Szemerényi, *Richtungen der modernen Sprachwissenschaft* (Heidelberg: Winter, 1971), vol. 1.

2. Michel Serres, "Platonic Dialogue," in *Hermes: Literature, Science, Philosophy*, ed. Josué V. Harari and David F. Bell (Baltimore: Johns Hopkins University Press, 1992), 66.

3. Ibid., 69.

4. See Michel Serres, *The Parasite*, trans. Lawrence R. Schehr (Baltimore: Johns Hopkins University Press, 1982), 53.

5. Ibid., 13 (translation emended).

6. For instance, fourteenth-century French legal experts discovered that suppressing highway robbery would profit the king. Though roads were not royal property they were *hors du commerce*, which enabled the king to claim a protective function. Highway robbery became a means for extending the monarch's territorial power beyond his domain—roads acted as swaths into territories that were ruled over by the local nobility. See Paul Alliès, *L'invention du territoire* (Grenoble: Presses Universitaires de Grenoble, 1960), 157.

7. Serres, *The Parasite*, 63.

8. Serres, "Platonic Dialogue," 67 (emphasis in the text).

9. For details see Roman Jakobson, "Linguistics and Poetics," in *Style in Language*, ed. Thomas Sebeok (Cambridge: MIT Press), 130–44.

10. Bruce Clarke, "Constructing the Subjectivity of the Quasi-Object: Serres through Latour," lecture given at "Constructions of the Self: The Poetics of Subjectivity," University of South Carolina, April 1999.

11. "There can be no doubt that we have here a new type of linguistic use—phatic communion I am tempted to call it, actuated by the demon of terminological invention—a type of speech in which ties of union are created by a mere exchange of words." Bronislaw Malinowski, "The Problem of Meaning in Primitive Languages," supplement to *The Meaning of Meaning: A Study of the Influence of Language upon Thought and of the Science of Symbolism*, by C. K. Ogden and I. A. Richards (London: Routledge and Kegan Paul, 1949), 315.

12. Malinowski, "The Problem of Meaning," 315.

13. Ibid., 314.

14. The nexus between the communion of bread and the communion of words in Malinowski is analyzed in greater detail in the following chapter, "Eating Animals—Eating Gods—Eating Man." See the "Ambivalence of the Tongue" section of this chapter.

15. Clarke, "Constructing the Subjectivity of the Quasi-Object."

16. Theodor Mommsen, ed., *Res gestae Divi Augusti* (Berlin: apud Weidmannos, 1865), 4.

17. Quoted in Jean Leclercq, "Saint Bernard et ses sécrétaries," *Révue bénédictine* 61 (1951): 208–09.

18. Andreas Schottus, "Ampliss: Viro Augerio Busbequio Exlegato Byzantino, & supremo Curiae Isabellae Praefecto" (Dedication), in Sextus Aurelius Victor, *De vita et moribus imperatorum romanorum*, ed. Andreas Schottus (Antwerp: Ex officina Christophori Plantini, 1579), 6.

19. *The Turkish Letters of Ogier Ghiselin de Busbecq*, trans. Edward Seymour Forster (Oxford: Oxford University Press, 1968), 49.

20. Ibid., 55.

21. Ibid., 50.

22. Or Busbecq himself? Mommsen's edition of the *Res gestae* contains several versions of the inscription. The section titled "Exemplum Busbequinam" reproduces only one anonymous insertion: *desiderantur quinque lineae* (Mommsen, *Res gestae*, xiv). This seems to suggest that the remaining interpolations are the work of Schott.

23. Augustus, "Res gestae divi Augusti," in Aurelius Victor, *De vita et moribus imperatorum romanorum*, 70.

24. Ibid., 77.

25. See Brian Rotman, *Signifying Nothing: The Semiotics of Zero* (Stanford: Stanford University Press, 1993), 14–22, 28–46.

26. Norman Bryson, quoted in Rotman, *Signifying Nothing*, 32f.

27. Michel Foucault, *The Archeology of Knowledge*, trans. A. M. Sheridan Smith (New York: Pantheon, 1972), 138.

28. See Wolf Peter Klein and Marthe Grund, "Die Geschichte der Auslassungspunkte: Zu Entstehung, Form und Funktion der deutschen Interpunktion," *Zeitschrift für germanistische Linguisitk* 25, no. 1 (1997): 26.

29. Franz Kafka, letter of January 22 to 23, 1913, in Franz Kafka, *Letters to Felice*, ed. Erich Heller and Jürgen Born, trans. James Stern and Elisabeth Duckworth (New York: Schocken, 1973), 166. *Translator's note*: In line with this maritime reading, it is interesting to note that in German, "telephone receiver" is *Hörmuschel* (literally, "hearing shell").

30. See Gerhard Neumann, "Nachrichten vom 'Pontus': Das Problem der Kunst im Werk Franz Kafkas," in *Franz Kafka Symposion 1983*, ed. Wilhelm Emmrich and Bernd Goldmann (Mainz: Hase und Koehler: 1985), 194f.

31. Franz Kafka, *The Castle*, trans. Willa and Edwin Muir (London: Secker and Warburg, 1957), 95.

32. See Kafka, *Letters to Felice*, 158, and further, Rüdiger Campe, "Pronto! Telefonate und Telefonstimmen," in *Diskursanalysen I: Medien*, ed. Friedrich A. Kittler et al. (Opladen: Westdeutscher Verlag, 1987), 86. In a letter to Felice Bauer from January 17, 1913, Kafka mentions that he had just read an old set of *Die Gartenlaube*, a family magazine, from 1863. That set included an essay by Philipp Reis on the first telephone experiments.

33. See August Kraatz, *Maschinentelegraphen* (Braunschweig: Friedrich Vieweg, 1906).

34. In its original version, the text, which was published as part of the series *rot* under the title *vielleicht zunächst wirklich nur: der monolog der terry jo im mercy hospital* (Maybe at first really only: The monologue of Terry Jo in Mercy Hospital) consisted only of the monologue. For the radio version Ludwig Harig enlarged the script by adding the voices of the people who had participated in the murder of Terry Jo's family, the event that had left her adrift.

35. Claude Elwood Shannon, "A Mathematical Theory of Communication," in *Claude Elwood Shannon: Collected Papers*, ed. N. J. A. Sloan and Aaron D. Wyner (Piscataway, N.J.: IEEE Press, 1993), 15.

36. For further details see Shannon, "A Mathematical Theory of Communication," 14–15.

37. Max Bense and Ludwig Harig, "Der Monolog der Terry Jo," in *Neues Hörspiel: Texte, Partituren*, ed. Klaus Schöning (Frankfurt/M.: Suhrkamp, 1969), 59–61.

38. Max Bense, "Jabberwocky: Text und Theorie, Folgerungen zu einem Gedicht von Lewis Carroll," in *Radiotexte: Essays, Vorträge, Hörspiele*, ed. Caroline Walther and Elisabeth Walther (Heidelberg: Winter, 2000), 71.

39. Ibid.

40. Max Bense, *Einführung in die informationstheoretische Ästhetik: Grundlagen und Anwendungen in der Texttheorie* (Reinbek: Rowohlt, 1969), 20.

41. Ibid., 28.

42. Michel Serres, "The Origin of Language," in *Hermes: Literature, Science, Philosophy*, ed. Josué V. Harari and David F. Bell (Baltimore: Johns Hopkins University Press, 1992), 77.

43. Ibid., 78.

2. EATING ANIMALS — EATING GOD — EATING MAN: VARIATIONS ON THE LAST SUPPER, OR, THE CULTURAL TECHNIQUES OF COMMUNION

1. Immanuel Kant, "Eternal Peace," in *The Philosophy of Kant: Immanuel Kant's Moral and Political Writings*, ed. Carl J. Friedrich (New York: Modern Library, 1949), 446.

2. See Yuri Slezkine, *The Jewish Century* (Princeton: Princeton University Press, 2004), 12–14.

3. See Bernhard Kathan, *Zum Fressen gern: Zwischen Haustier und Schlachtvieh* (Berlin: Kadmos, 2004), 90.

4. Edward Wouk, "Dirk Bouts's Last Supper Altarpiece and the Sacrament van Mirakel at Louvain," *Immediations: The Research Journal of the Courtauld Institute of Art* 1, no. 2 (2005): 47.

5. Ibid., 44–47.

6. See Miri Rubin, *Gentile Tales: The Narrative Assault on Late Medieval Jews* (New Haven: Yale University Press, 1999).

7. Thomas Macho, "Tier," in *Vom Menschen: Handbuch Historische Anthropologie*, ed. Christoph Wulf (Weinheim-Basel: Beltz, 1997), 63 and passim.

8. Claude Lévi-Strauss, *The Savage Mind* (London: Weidenfeld and Nicolson, 1972), 224.

9. Elias Canetti, *Crowds and Power*, trans. Carol Stewart (New York: Viking, 1962), 357.

10. Umberto Eco, "Function and Sign: The Semiotics of Architecture," in *Rethinking Architecture: A Reader in Cultural Theory*, ed. Neil Leach (London: Routledge, 1997), 182.

11. Salcia Landmann, *Jesus und die Juden* (Munich: Herbig, 1987), quoted in Adolf Holl, "Das erste Letzte Abendmahl," in *Speisen, Schlemmen, Fasten: Eine Kulturgeschichte des Essens*, ed. Uwe Schultz (Frankfurt/M: Insel, 1993), 47–48.

12. See Holl, "Das erste Letzte Abendmahl," 48.

13. See Jochen Hörisch, *Brot und Wein: Die Poesie des Abendmahls* (Frankfurt/M: Suhrkamp, 1992), 58.

14. Heinrich von Kleist, *The Feud of the Schroffensteins*, trans. Mary J. and Lawrence M. Price, *Poet Lore* 27, no. 5 (1916), 457–72.

15. Further see Gerhard Neumann, "Hexenküche und Abendmahl: Die Sprache der Liebe im Werk Heinrich von Kleists," *Freiburger Universitätsblätter* 91 (1986): 9–31.

16. Quoted in Gerhard Neumann, "Tania Blixen: 'Babettes Gastmahl,'" in *Kulturthema Essen: Ansichten und Problemfelder*, ed. Alois Wierlacher, Gerhard Neumann, and Hans J. Teuteberg (Berlin: Akademie Verlag, 1993), 291.

17. Further see Georges Didi-Huberman, *La peinture incarnée* (Paris: Éditions de Minuit, 1985).

18. Roman Jakobson, "Linguistics and Poetics," in *Style in Language*, ed. Thomas Sebeok (Cambridge: MIT Press, 1960), 350–77.

19. Bronislaw Malinowski, "The Problem of Meaning in Primitive Languages," supplement to *The Meaning of Meaning: A Study on the Influence of Language upon Thought and of the Science of Symbolism*, by Charles. K. Ogden and Ivor A. Richards (London: Routledge, 1994), 463. All further page references, appearing in parentheses in the text, refer to this edition.

20. Quoted in Gerhard Neumann, "'Jede Nahrung ist ein Symbol': Umrisse einer Kulturwissenschaft des Essens," in *Kulturthema Essen*, ed. Wierlacher, Neumann, and Teuteberg, 387.

21. Werner Hamacher, *Pleroma: Reading in Hegel; The Genesis and Structure of a Dialectical Hermeneutics in Hegel*, trans. Nicholas Walker and Simon Jarvis (London: Athlone, 1998), 101.

22. Ibid., 102.

23. G. W. F. Hegel, "The Spirit of Christianity," in *Early Theological Writings*, trans. T. M. Knox (Philadelphia: University of Pennsylvania Press, 1971), 249.

24. Immanuel Kant, *Religion within the Bounds of Bare Reason*, trans. Werner Pluhar (Indianapolis: Hackett, 2009), 197. All further page references, which appear in parentheses following quotations in the text, are to this edition.

25. "[U]t 'est' pro 'significat' accipiamus [necesse est]." "Der Bericht Hedios," in *Das Marburger Religionsgespräch 1529*, ed. Gerhard May (Gütersloh: Mohn, 1970), 21.

26. See "Bericht Osianders an den Nürnberger Rat," in *Das Marburger Religionsgespräch 1529*, ed. May, 54.

27. See entry by Otto Hiltbrunner, Denys Gorce, and Hans Wehr, "Gastfreundschaft" ["Hospitality"] in Theodor Klausman et al., eds, *Reallexikon für Antike und Christentum*, vol. 7 (Stuttgart: Hiersemann, 1972), cl. 1080. Further see Otto Hiltbrunner, *Gastfreundschaft in der Antike und im frühen Christentum* (Darmstadt: Wissenschaftliche Buchgesellschaft, 2005).

28. Hamacher, *Pleroma:Reading in Hegel*, 103.

29. "Der Bericht Hedios," 20.

30. Jean-Jacques Rousseau, *Confessions*, trans. Angela Scholar (Oxford: Oxford University Press, 2000), 106.

31. From the play's final scene:

Penthesilea: You mean that I—? You claim I—him—
My dogs and I together—? You say hands as small
as these—? And a mouth like this, with love-swelled
lips—? Shaped for such a different service
than to—! Helping each other to go at
it, avidly, the mouth and then the hand,
the hand and then the mouth—?

  . . .

Didn't kiss him, no? Really tore his flesh
to shreds? . . .
An error, then,
I see, A kiss, a bite—how cheek by jowl
they are, and when you love straight from the heart
the greedy mouth so easily mistakes
one for the other.

Heinrich von Kleist, *Penthesilea*, in *Five Plays*, trans. Martin Greenberg (New Haven: Yale University Press, 1988), 264f.

32. See Jean-Jacques Rousseau, *Politics and the Arts: Letter to M. d'Alembert on the Theatre*, trans. Allen Bloom (Glencoe: Free Press, 1960), 123–37.

33. Norbert Elias, *The Civilizing Process*, vol 1: *The History of Manners*, trans. Edmund Jephcott (New York: Urizen, 1978), 89.

34. Ibid., 87.

35. See Hans Ottomeyer, "Tischgerät und Tafelbräuche: Die Kunstgeschichte als Beitrag zur Kulturforschung des Essens," in *Kulturthema Essen: Ansichten und Problemfelder*, ed. Alois Wierlacher, Gerhard Neumann, and Hans J. Teuteberg (Berlin: Akademie Verlag, 1993), 178. See also Gert von Paczensky and Anna Dünnebier, *Leere Töpfe, volle Töpfe: Die Kulturgeschichte des Essens und Trinkens* (Munich: Knaus, 1994).

36. See Claus Dieter Rath, *Reste der Tafelrunde: Das Abenteuer der Eßkultur* (Reinbek: Rowohlt, 1984), 216.

37. See Gisèle Harrus-Révidi, *Die Kunst des Geniessens: Esskultur und Lebenslust*, trans. Renate Sandner and Thorsten Schmidt (Düsseldorf and Zürich: Artemis und Winkler, 1996), 129f.

38. See Jacques Derrida, *Rogues: Two Essays on Reason* (Stanford: Stanford University Press, 2005).

39. Thomas Pynchon, "Entropy," in *Slow Learner* (Boston and Toronto: Little, Brown, 1984), 79–98.

40. Thomas Pynchon, "Mortality and Mercy in Vienna," *Epoch* 9, no. 4 (1959): 195–213. All further page references, appearing in parentheses in the text, are to this edition.

41. See Hiltbrunner, Gorce, Wehr, "Gastfreundschaft."

42. Joseph Conrad, *Heart of Darkness* (London and New York: Penguin, 1985), 111.

3. *PARLÊTRES*: THE CULTURAL TECHNIQUES OF ANTHROPOLOGICAL DIFFERENCE

1. See Gilles Deleuze and Felix Guattari, *A Thousand Plateaus: Capitalism and Schizophrenia,* trans. Brian Massumi (Minneapolis: University of Minnesota Press, 1987), 240–41.

2. Lacan introduced the neologism *parlêtre* in 1974 to indicate that "carnal being" which is "haunted by the word." See Jacques Lacan, *Le triomphe de la religion* (Paris:

Éditions du Seuil, 2005), 90, and "7ème Congrès de l'École freudienne de Paris à Rome," *Lettres de l'École freudienne*, no. 16 (1975): 177–203. The latter has been published in English as "Press Conference by Doctor Jacques Lacan at the French Cultural Center, Rome, 29-October-1974," trans. Richard G. Klein, http://www.freud2lacan.com/docs /lacan_press.pdf (accessed 10 July 2012).

3. See Horst Waldemar Janson, *Apes and Ape Lore in the Middle Ages and the Renaissance* (London: Warburg Institute, 1952), 287–325.

4. Aristotle, *The Politics* [1253a7–10], trans. J. A. Sinclair (Harmondsworth: Penguin, 1992), 60.

5. René Descartes, *Discourse on Method*, trans. Richard Kennington (Newburyport, Mass.: R. Pullins, 2007), 46. All subsequent page references appear in parentheses in the text and are to this edition.

6. See Friedrich A. Kittler. "Autorschaft und Liebe," in *Austreibung des Geistes aus den Geisteswissenschafte: Programme des Poststrukturalismus*, ed. Friedrich A. Kittler (Paderborn: Schöningh, 1980), 157.

7. See Rüdiger Campe, *Affekt und Ausdruck: Zur Umwandlung der literarischen Rede im 17. und 18. Jahrhundert* (Tübingen: M. Niemeyer, 1990), 79–80.

8. *The Natural History of Pliny*, trans. John Bostock and H. T. Riley (London: George Bell, 1890), 2:523–24 (translation emended).

9. Aristotle, *Historia Animalium*, trans. D'Arcy Wentworth Thompson, vol. 4 of *The Works of Aristotle*, ed. J. A. Smith and W. D. Ross (Oxford: Clarendon Press, 1910), 597b.

10. E.g., see Gabriele Salci, *Natura morta con strumenti musicali, fiori, frutta e verdura*, painting, reproduced in *Stilleben: Die italienischen, spanischen und französischen Meister*, ed. Claus Grimm (Stuttgart: Belser, 1995), 187. Salci depicts a parrot with a couple of wine glasses in front of a sheet of music.

11. Pliny, *Natural History*, 2:523.

12. Dante Alighieri, *De vulgari eloquentia*, ed. and trans. Steven Botterill (Cambridge and New York: Cambridge University Press, 1996), 7.

13. Azzo, *Summae Institutionum* 4, 50. Quoted in Donald E. Queller, *The Office of Ambassador in the Middle Ages* (Princeton: Princeton University Press, 1967), 7.

14. "Ubi nos presentes esse non possumus, nostra per eum cui precipimus, representetur auctoritas." Quoted in Franz Wasner, "Fifteenth-Century Texts on the Ceremonial of the Papal 'Legatus a latere,'" *Traditio: Studies in Ancient and Medieval History, Thought and Religion* 14 (1958): 300.

15. Garrett Mattingly, *Renaissance Diplomacy*, quoted in Queller, *Office of the Ambassador*, 9.

16. Baldus, *Commentaria*, ad C. 4, Si quis alteri vel sibi, 4, 17. Quoted in Queller, 9.

17. "Et nuncii aliique, quorum opera in negociis absentium utimur, nihil aliud sunt, quam quasi organa sive picae (ut alibi dicit Bartholus) quibus absentium inter se sermones conferuntur." Conradus Brunus, "De legationibus libri quinque: Cunctis in repub. versantibus, aut quolibet magistratu fungentibus perutiles, & lectu iucundi," in *D. Conradi Bruni iureconsulti opera tria* (Mainz: Francis Behem, 1548), 2.

18. See Heinrich Bosse, "Dichter kann man nicht bilden: Zur Veränderung der Schulrhetorik nach 1770," *Jahrbuch für Internationale Germanistik* 10 (1978): 80–125. Further see Bosse, "'Die Schüler müssen selbst schreiben lernen,' oder, Die Einrichtung der Schiefertafel," in *Schreiben—Schreiben lernen: Rolf Sanner zum 65. Geburtstag*, ed. Dietrich Boneke and Norbert Hopster (Tübingen: Narr, 1985), 164–99.

19. Johann Gottfried von Herder, "Vitae non scholae discendum," in *Sämtliche Werke*, ed. Bernhard Suphan (Berlin: Weidmann,1877–1913), 30:267.

20. *Translator's note:* This particular phrasing is a deliberate allusion to §459 in Hegel's *Encyclopedia of the Philosophical Sciences*, in which the art of learning to read and write is said to contribute and "give stability and independence to the inward realm of mental life."

21. Johann Gottfried von Herder, "Von der Ausbildung der Rede und Sprache in Kindern und Jünglingen," in *Sämtliche Werke*, ed. Bernhard Suphan (Berlin: Weidmann, 1877–1913), 30:217f.

22. In contrast there were times when language was fully aligned with creaturely dialects. As Kittler points out in *Discourse Networks,1800/1900*, trans. Michael Metteer and Chris Cullens (Stanford: Stanford University Press, 1990), during the Reformation some German primers introduced consonants and consonant blends in a thoroughly beastly fashion: "[Jacob] Grüssbeutel's *Little Voice Book* presented *ss* as a hissing snake, *pf* as a snarling cat being barked at by dogs. Peter Jordan's *Lay Book* gave these rules of pronunciation: 'The *l* as the ox lows. The *m* as the cow moos. The *r* as the dog growls. The *s* as the young doves whistle and coo.' . . . The sixteenth-century conception of language directed children toward the many languages of creation" (38–39). The source of these animal voice catalogs was Valentin Ickelsamer's *Teütsch Grammatica* (German Grammar) of 1533.

23. Herder, "Treatise on the Origin of Language (1772)," in Herder, *Philosophical Writings*, trans. and ed. Michael N. Forster (Cambridge: Cambridge University Press, 2002), 65. All further page references appear in parentheses in the text and are to this edition.

24. *Translator's note:* For the following it is important to note that Herder uses neither the German word for ram (*Widder*) nor an unequivocal description such as *männliches Schaf* ("male sheep"), but the quaint and potentially ambiguous *Schaafmann*. It literally means "sheep-man"; and as several tongue-in-cheek commentators including Siegert have pointed out, it can also be read as referring to a shepherd impelled by instinct to become very closely acquainted with members of his flock.

25. See Thomas Macho, "Tier," in *Vom Menschen: Handbuch Historische Anthropologie*, ed. Christoph Wulf (Weinheim: Beltz, 1997), 71.

26. See Burchard Brentjes, *Die Erfindung des Haustieres*, 3rd ed. (Leipzig: Urania-Verlag, 1986), 28.

27. See Kittler, *Discourse Networks*, 39.

28. Johann Heinrich Pestalozzi, "Über den Sinn des Gehörs in Hinsicht auf Menschenbildung durch Ton und Sprache," in *Ausgewählte Schriften*, ed. Wilhelm Flitner (Frankfurt/M.: Ullstein, 1983), 248.

29. Gustave Flaubert, *A Simple Heart*, trans. Charlotte Mandell (Hoboken, N.J.: Melville House, 2004), 3. All further page references, appearing in parentheses in the text, are to this edition.

30. Julian Barnes, *Flaubert's Parrot* (New York: Knopf, 1985), 17.

31. See David Fraser, "Joseph Wright of Derby et la Lunar Society: Essai sur les rapports de l'artiste avec la science et l'industrie," in *Joseph Wright of Derby 1734–1797*, exhibition catalog, ed. Judy Egerton (Paris: Réunion des Musées Nationaux, 1990), 24.

32. See William Schupbach, "A Select Iconography of Animal Experiment," in *Vivisection in Historical Perspective*, ed. Nicolaas Rupke (London: Croom Helm, 1987), 347.

33. Werner Busch, *Joseph Wright of Derby: Das Experiment mit der Luftpump, eine Heilige Allianz zwischen Wissenschaft und Religion* (Frankfurt/M: Fischer Taschenbuch, 1986).

34. Such technological implementations of spiritual channels, usually the domain of enlightened profanation, can already be found in the Middle Ages. The north portal of the Lady Chapel at Würzburg depicts a pliable tube that connects God's mouth with Mary's ear, upon which a tiny Jesus glides into his future mother's ear. It is a spiritual channel that allows for the undistorted transmission of God's breath, the divine logos. See Manfred Schneider, *Liebe und Betrug: Die Sprachen des Verlangens* (Munich and Vienna: Carl Hanser Verlag, 1992), 248.

35. See Gustave Flaubert, *Carnets de travail*, critical and genetic edition, ed. Pierre-Marc de Biasi (Paris: Balland, 1988): "Wright: Expérience de la machine pneumatique. Effet de nuit. Deux amoureux dans un coin, charmants. Le vieux 'à longs cheveux' qui montre l'oiseau sous le verre. Petite fille qui pleure. Charmant de naïveté et de profondeur" (350).

36. For the modern interest in glossolalia in the sciences and pneumatic revivalism since the seventeenth century, see Thomas Macho, "Glossolalie in der Theologie," in *Zwischen Rauschen und Offenbarung: Zur Kultur- und Mediengeschichte der Stimme*, ed. Friedrich Kittler, Thomas Macho, and Sigrid Weigel (Berlin: Akademie Verlag, 2002), 3–17.

37. In late July 1876 Flaubert wrote to Madame Charles Roger des Genettes (and similar sentiments are to be found in a letter of July 28, 1876, to Madame Jean-Baptiste Brainne): "For over a month I have had a stuffed parrot perched on my desk in order to 'copy' nature. Its presence has become fatiguing. Nonetheless I keep him in order to fill my soul with parrot."

38. Barnes, *Flaubert's Parrot*, 57.

39. Letter to Madame des Genettes (January 24, 1880), in *The Letters of Gustave Flaubert, 1857–1880*, ed. and trans. Francis Steegmuller (Cambridge: Harvard University Press, 1982), 263.

40. See Barnes, *Flaubert's Parrot*, 57–58.

41. Albert Thibaudet, *Gustave Flaubert, 1821–1880: Sa Vie—Ses Romans—Son Style* (Paris: Plon-Nourrit et Cie, 1922), 236.

42. Gustave Flaubert, *Correspondance*, vol. 2: *1847–1852* (Paris: Louis Conard, 1926), 185.

4. MEDUSAS OF THE WESTERN PACIFIC: THE CULTURAL TECHNIQUES OF SEAFARING

1. Michel Foucault, "Of Other Spaces," in *The Visual Culture Reader*, ed. Nicholas Mirzoeff (London and New York: Routledge, 1998), 244.

2. Michel Serres, "Turner traduit Carnot," in *Hermes III: La Traduction* (Paris: Éditions de Minuit, 1974), 237. ("A ship is always a perfect summary of space as it is.")

3. See the introduction to this volume.

4. Foucault, "Of Other Spaces," 244.

5. Ibid., 242.

6. Hans Blumenberg, *Shipwreck with Spectator: Paradigm of a Metaphor for Existence*, trans. Steven Rendall (Cambridge: MIT Press, 1997), 7.

7. Sophocles, *The Three Theban Plays: Antigone, Oedipus the King, Oedipus at Colonus*, trans. Robert Fagles (Harmondsworth: Penguin, 1982), 76.

8. Plato, *Timaeus* [52b], trans. Donald J. Zeyl (Indianapolis: Hackett Publishing, 2000), 41–42.

9. Martin Heidegger, *Introduction to Metaphysics*, trans. Ralph Manheim (New Haven: Yale University Press, 1959), 66.

10. Christina Vagt, *Geschickte Sprünge: Physik und Medium bei Martin Heidegger* (Zürich: Diaphanes, 2010), 167.

11. Hesiod, *"Theogony" and "Works and Days,"* trans. M. L. West (Oxford: Oxford University Press, 1989), 56–57.

12. Bronislaw Malinowski, *Argonauts of the Western Pacific: An Account of Native Enterprise and Adventure in the Archipelagos of Melanesian New Guinea*, with introduction by James G. Fraser (New York: E.P. Dutton, 1961). All subsequent page references appearing in parentheses in the text are to this edition.

13. "This detailed geographical indexing is actually a mapping of actual sea routes for kula and other expeditions that are in current use. The mythic routes then code knowledge of actual sailing routes, though the mythic voyages are much longer than those usually undertaken." S. J. Tambiah, "On Flying Witches and Flying Canoes: The Coding of Male and Female Values," in *The Kula: New Perspectives on Massim Exchange*, ed. Jerry W. Leach and Edmund Leach (Cambridge, New York, and Melbourne: Cambridge University Press, 1983), 187.

14. The notion of a gaping depth also underlies the word *chaos* (from Greek *kaino* = "gaoe," "yawn"). Apollonius Rhodius described the night voyage of the Argonauts across the "shroud" of the Cretan sea as a passage through *melan chaos*, "black chaos" (*Argonautica*, trans. William H. Race [Cambridge: Harvard University Press, 2008], 465). According to Hesiod, chaos was created first (*"Theogony" and "Works and Days,"* 6). Ovid identifies Hesiod's chaos with the old Stoic notion of matter without quality (see Gernot Böhme and Hartmut Böhme, *Feuer, Wasser, Erde, Luft* [Munich: Beck, 1996], 34).

15. "Perhaps the most extraordinary belief in this connection is that the *tokwalu*, the carved human figures on the prow boards, the *guwaya*, the semihuman effigy on the

mast top, as well as the canoe ribs would 'eat' the drowning men if not magically 'treated'" (Malinowski, *Argonauts*, 246).

16. Thomas Hauschild, *Der böse Blick: Ideengeschichtliche und sozialpsychologische Untersuchungen*, 2nd ed. (Berlin: Verlag Mensch und Leben, 1982), 192.

17. Sigmund Freud, "From the History of an Infantile Neurosis," in *The Standard Edition of the Complete Psychological Works of Sigmund Freud*, vol. 17, *An Infantile Neurosis and Other Works*, ed. James Strachey (London: Hogarth, 1964), 84.

18. See Jacques Lacan, *The Seminar of Jacques Lacan, Book 1: The Four Fundamental Concepts of Psychoanalysis*, ed. Jacques-Alain Miller, trans. Alan Sheridan (New York: Norton, 1998), 62.

19. Slavoj Žižek, "'I Hear You with My Eyes', or, The Invisible Master," in *Gaze and Voice as Love Objects*, ed. Renata Salecl and Slavoj Žižek (Durham, N.C.: Duke University Press, 1996), 91.

20. Lacan, *Seminar, Book 11*, 72.

21. Ibid., 95.

22. Ibid.

23. See Jacques Lacan, *Le séminaire de Jacques Lacan, livre X: L'angoisse*, ed. Jacques-Alain Miller (Paris: Seuil, 2004), 293.

24. Edmund R. Leach, "A Trobriand Medusa?" *Man* 54 (July 1954): 103–05. See also Richard F. Salisbury, "Trobriand Medusa?" *Man* 59 (March 1959): 50–51. Thanks to Thomas Hauschild for pointing out Leach's paper.

25. See Robert Graves, *The Greek Myths*, rev. ed. (Harmondsworth: Penguin, 1986), 1:241.

26. See Josef Floren, *Studien zur Typologie des Gorgoneion* (Munich: Aschendorff, 1977), 12.

27. A spell used to suspend the eating taboo reveals that the creeper (*wayugo*) used to lash together the canoe is also endowed with a hostile gaze: "I sprinkle thy eye, O *kudayuri* creeper, so that our crew might eat" (Malinowski, *Argonauts*, 229). The name of the creeper refers once again to the myth of the flying canoe, probably because breaking the eating taboo would slow down the canoe.

28. Like numerous other vases, the famous stamnos depicting the ship of Odysseus passing the sirens (from around 475 B.C.E.) also shows the ship sporting an eye on the prow. See John S. Morrison and Roderick T. Williams, *Greek Oared Ships, 900–322 B.C.* (Cambridge: Cambridge University Press, 1968), 114/21e. For further examples of ships' eyes see figures 9a and c, 10a and c, 13, 14c-g, 15b, 17a and d, 18a and b, and 20d. When compared with archeological findings, it appears that the eyes on the vases were painted oversize.

29. See Troy Joseph Nowak, "Archaeological Evidence for Ship Eyes: An Analysis of their Form and Function," master's thesis, Texas A&M University, 2006, 156–70.

30. See Nowak, "Archaeological Evidence," 81–115.

31. See Assaf Yasur-Landau, "On Birds and Dragons: A Note on the Sea Peoples and Mycenaean Ships," in *Pax Hethitica: Studies on the Hittites and Their Neighbours in Honour of Itamar Singer*, ed. Yoram Cohen, Amir Gilan, and Jared L. Miller (Wiesbaden:

Harrassowitz, 2010), 402. A seal in Knossos depicts such a sea monster, sporting the head of a dog, being driven off by a hero on board the ship.

32. See Fernand Benoit, "Jas d'ancre à téte de Meduse," *Revue archéologique* 6, no. 37 (1951): 223–28, especially 226.

33. See Hauschild, *Der böse Blick*, 192.

34. See Alison Luchs, *The Mermaids of Venice: Fantastic Sea Creatures in Venetian Renaissance Art* (London: Harvey Miller Publishers, 2010), 14–16.

35. See Károly Marót, *Die Anfänge der griechischen Literatur: Vorfragen* (Budapest: Verlag der Ungarischen Akademie der Wissenschaften, 1960), 108. The siren adventure may already have played a central part in pre-Homeric Argonaut songs. According to Marót the sirens originally were "muses of magic song" and probably originate in Phoenician maritime folklore (Marót, 147).

36. See Hauschild, *Der böse Blick*, 195. The close link between the evil eye and the doubling of a woman into good original and evil doppelganger also shapes the legends of Saint Lucy, though they are no longer associated with the maritime context. According to a thirteenth-century Italian version, Lucy tore out her beautiful eyes and sent them to her spellbound pagan admirer. A latter version has Mary give her new, even more beautiful eyes. Removed eyes enjoyed special popularity in fifteenth-century Italian art. Carlo Crivelli's *Santa Lucia* (around 1476, and now in the National Gallery in London) depicts Lucia carrying her eyes on a plate and wearing a ruby around her neck whose light reflexes have something of a gaze to them. In German-Slavic regions we encounter St. Lucy's pagan opposite, Lucia the Dark, also called Perchta or Bercht, who in the middle of the night threatens to poke out the eyes of disobedient children. Further see Leopold Kretzenbacher, *Santa Lucia und die Lutzelfrau: Volksglaube und Hochreligion im Spannungsfeld Mittel- und Südeuropas* (Munich: Oldenbourg, 1959), 14–15, 21, and passim.

37. On the design of Trobriand war shields and their possible relation to similar shields on the Solomon Islands, see Philip C. Gifford, "Trait Origins in Trobriand War-Shields: The Uncommon Selection of an Image Cluster," *Anthropological Papers of the American Museum of Natural History* 79 (1996). Gifford supports Leach's reading of the shield design by interpreting the patterns formed by the snake bands in the lower part of the image as a vulva (8). Similar patterns, incidentally, can be found on the steering paddles (*vivoyu*) of Trobriand canoes.

38. See Tambiah, "On Flying Witches and Flying Canoes," 188.

39. Of course there are exceptions, such as the strange painting by Hans Saverij I, *Sea Monsters* (1626, Amsterdam, private collection), that depicts several monsters and fish gazing with wide open eyes at the observer. See Margarita Russell, *Visions of the Sea: Hendrick C. Vroom and the Origins of Dutch Marine Painting* (Leiden: Brill and Leiden University Press, 1983), 50.

40. See Hans Belting, *Bild-Anthropologie: Entwürfe für eine Bildwissenschaft* (Munich: Wilhelm Fink Verlag, 2001), 144–45.

41. In a later response to criticisms of his interpretation of Trobriand shield paintings, Leach classified the flying witches as "figments of a paranoid imagination," thereby

implying that the sea of the Trobriands is the space of psychosis. See Edmund Leach, "'A Trobriand Medusa?' A Reply to Dr. Berndt," *Man* 58 (May 1958), 79.

5. *PASAJEROS A INDIAS*: REGISTERS AND BIOGRAPHICAL WRITING AS CULTURAL TECHNIQUES OF SUBJECT CONSTITUTION (SPAIN, SIXTEENTH CENTURY)

1. When the Casa de la Contratación de las Indias in Seville was founded in 1503, its initial function was to control the trade between Spain and its new overseas colonies. But only a few years later it also took on scientific functions (supervising maps, mapmakers, and their instruments as well as training and examining pilots). This accumulation of administrative, educational, and nautical functions turned the Casa de la Contratación into "one of the most important centers of applied sciences in the sixteenth century" (José Maria López de Pinero, *El arte de navegar en la España del Rinacimiento* [Barcelona: Editoreal Labor, 1986], 127). In addition to its scientific tasks, the trade house increasingly had to attend to juridical functions as well. The so-called *juezes-oficiales* of the Casa exercised complete civil and partial penal jurisdiction over captans, pilots, crews, and passengers. From a certain date on, the Casa had its own remand prison.

2. John Torpey, *The Invention of the Passport: Surveillance, Citizenship and the State* (Cambridge: Cambridge University Press, 2000), 1.

3. Ibid., 2.

4. See Valentin Groebner, *Who Are You? Identification, Deception, and Surveillance in Early Modern Europe*, trans. Mark Kyburz and John Peck (New York: Zone Books, 2007).

5. See Michel Foucault, "Society Must Be Defended," in *Ethics: Subjectivity and Truth*, ed. Paul Rabinow (New York: New Press, 1997), 59–65.

6. Archivo General de Indias (hereafter abbreviated AGI), Seville, Contratación, 5230, numero 1, ramo 2, fol. 1. Catalina succeeded: see AGI, Pasajeros, libro 6, expediente 5176.

7. See *Recopilación de leyes de los reynos de las Indias* (Madrid, 1681; reprinted, Mexico City: M. A. Porrúa, 1987), libro 9, título 26, ley 25: "Algunas Mugeres casadas, que tienen en las Indias sus maridos, piden licencia para passar á aquellas partes, y hazer vida mardiable con ellos, y muestran, que las envian á llamar, porq se les manda en las Indias, que vengan poi sus mugeres" (November 1554 and July 1555). See also Joseph de Veitia Linage, *Norte de la Contratación de las Indias occidentales* (Seville: Blas, 1672), p. 309: "Por otro capítulo de la . . . carta de 10. Mayo de 1546 . . . se dá facultad para que á la muger casada que tuviere su marido en Indias, se le dé licencia, no solamente á ella, sino á vn deudo suyo."

8. See *Recopilación*, libro 9, título 26, ley 25, fol. 5r: "Mandamos al Presidente, y Iuezes de la Casa, que á las mugeres, que huviere de esta calidad . . . dexen passar, aunque no tengan licencia nuestra." With a license issued by the judges of the Casa, merchants were allowed to travel to the Indies for three years even if they were married; see *Recopilación*, libro 9, título 26, ley 29. The same was true for the Factores de Mercaderes; see *Recopilación*, libro 9, título 26, ley 37.

9. Quoted in Veitia Linage, *Norte de la Contratación*, 304:

Que de allí adelante no se consintiessen los Juezes Oficiales que passassen á ninguna parte de las Indias passagero alguno . . . , sin que levassen, y presentassen ante ellos informaciones hechas en sus tierras, y naturalezas . . . por donde constasse si son casados, ó solteros, y las señas, y edad que tienen, y que no son las nuevamente convertidos á nuestra Santa Fé Catolica de Moro, ó de Judio, ni hijo suyo, ni reconciliado, ni hijos, ni nietos de personas que publicamente huviere traído Sanbenito, ni hijos, ni nietos de quemados, ó condenados por Hereges por el delito de la heretica pravedad por linea masculina, ni femenina, y con aprovacion de la justicia de la Ciudad, Villa, ó Lugar donde la tal informacion se hiziere, en que se declare como la persona que assi dá la tal informacion, es libre, ó casado. (April 5, 1552)

10. Michel Foucault, *Discipline and Punish: The Birth of the Prison*, trans. Alan Sheridan (New York: Vintage, 1979), 191.

11. The significance of this question is discussed below.

12. AGI, Contratación, 5221, numero 1, ramo 3/1, fol. 3r: ". . . es muy quieto e pacifico e no ha sido ni es alborotador de pueblos ni zizañador antes es de buena vida y forma y muy bien jureinado e de buenas costumbres e quito de quistiones e ruydos e alborotos."

13. AGI, Contratación, 5221, numero 1, ramo 3/1, fol. 3r.

14. AGI, Contratación, 5220, numero 1, ramo 12/1; 8-10-1563 / Alvaro Rodriguez de Mendaña, fol. 4r.

15. AGI, Contratación, 5220, numero 1, ramo 7, fol. 1r.

16. AGI, Contratación, 5537, libro 1, fol. 4v:

El Principe:

Oficiales del emperador Rey. nr senor q[ue] residis en la ciudad de seuilla en la casa de la contratación de las yndias anos se ha fecho Relacion que muchos de los pasajeros [son] personas que conforme a lo que por nos esta mandado en las licencias q[ue] de nos lleuan puedan pasar a las yndias . . . que van a esa casa a dar las ynformaciones de si son casados, o no, o delos de mas que son obligados de dar la presentan testigos falsos para prouar lo que quieren ellos cerca desto de donde viene que muchos q[ue] son casados dan ynformacion que son libres e se hizen otras fraudes.

17. AGI, Contratación, 5220, numero 1, ramo 12/1, fol. 2r.

18. Real Cedula, 17 October 1544, repeated on 4 September 1549. Quoted in Norman F. Martin, *Los vagabundos en la Nueva Espana, siglo XVI* (Mexico City: Jus, 1957), 30.

19. "Ninguno se aplica a seruir ni trauajar, ni quieren, y andan las plaças y calles lleas de mugeres baldías y de hombres bagamundos perdidos y perdidas." Quoted in Martin, *Los vagabundos*, 88.

20. Hendrik Bolkestein and Adolf Kalsbach, "Armut I," in Theodor Klausman et al., eds., *Reallexikon für Antike und Christentum*, vol. 1 (Stuttgart: Hiersemann, 1950), cl. 698.

21. See Bronislaw Geremek, *Poverty: A History*, trans. Agnieszka Kolakowska (Oxford: Blackwell, 1994), 19.

22. Jesus decreed that perfection requires one to give up all possessions (see Matthew 19:21–24).

23. See Jean-Pierre Gutton, *La société et les pauvres en Europe, XVIe–XVIIIe siècles* (Paris: Presses Universitaires de France, 1974), 94.

24. Ibid.

25. See Michele Maccarrone, *Vicarius Christi: Storia del titolo papale*, Lateranum, new series 18 (Rome: Facultas Theologica Pontificii Athenaei Lateranensis, 1952), 119 sq. See also Ernst H. Kantorowicz, *The King's Two Bodies. A Study in Mediaeval Political Theology* (Princeton: Princeton University Press, 1957), 90–91.

26. Gremek, *Poverty*, 20.

27. See Florian Oberhuber, "Der Vagabund," in *Grenzverletzer: Von Schmugglern, Spionen und anderen subversiven Gestalten*, ed. Eva Horn, Stefan Kaufmann, and Ulrich Bröckling (Berlin: Kadmos, 2002), 61. Oberhuber's account focuses on the eighteenth century.

28. See Real Academia de la Historia, eds., *Las siete Partidas del Rey Don Alfonso el Sabio, cotejadas con varios codices antiguos*, vol. 1 (Madrid: Imprenta Real, 1807), partida 2, título 5, ley 40, and partida 2, título 20, ley 4.

29. See Pamphilus Gengenbach, *Von der falschen Betler buberey [Liber vagatorum]*, with introduction by Martin Luther (Wittemberg and Magdeburg: Öttinger, 1528).

30. Christobal Pérez de Herrera, *Discurso del amparo de los legítimos pobres y reducción de los fingidos y de la fundación de los albergues destos Reynos y amparo de la milicia dellos* (1598). Quoted in César Real Ramos, " 'Fingierte Armut' als Obsession und die Geburt des auktorialen Erzählers in der Picaresca," in *Der Ursprung von Literatur: Medien, Rollen, Kommunikationssituationen zwischen 1450 und 1650*, ed. Gisela Smolka-Koerdt, Peter M. Spangenberg, and Dagmar Tillmann-Bartylla (Munich: Wilhelm Fink Verlag, 1988), 179. Compare Matthew 19:29: "And everyone who has left houses or brothers or sisters or father or mother or wife or children or lands, for My name's sake, shall receive a hundredfold, and inherit eternal life."

31. AGI, Contratacion, 5221, numero 1, ramo 3/1, fol. 6v.:

Yo Antonio Velez escribano publico de su Real mag. en la su corte, Reinos e senorios e un del numero de la villa de Almacan doy fee e verdadero testimonio a todos los senores que la presente vieren e leyere.. como el dicho Juan de Ortega scribano de quien va signada esta Informacion e probanca: es scribano de su Real Mag.d e del numero desta dicha villa y scribano fiel y legal en su officio . . . Y ansimesmo doy fee quel dicho Auaro de soto de quien va firmada, es Alcalde hordinario dela dicha villa de Almacan y conozco ser su firma por q[ue] muchas firmas tengo yo en mis scripturas que son como la que aqui firmo. Fecho en la villa de Almacan a tres días de otubre, de mill y qui[nient]os y sesenta y tres años.

En este testimonio de verdad

Ant[oni]o velez, scribano

32. See ibid., fol. 6v.

33. Ibid.

34. See Jacques Derrida. "Signature Event Context," in *Limited Inc* (Evanston, Ill.: Northwestern University Press, 1988), 17–19.

35. Ibid., 17.

6. (NOT) IN PLACE: THE GRID, OR, CULTURAL TECHNIQUES OF RULING SPACES

1. Xenophon, *"Memorabilia" and "Oeconomicus,"* trans. E. C. Marchant (Cambridge: Harvard University Press, 1938), 439.

2. Ibid.

3. See Gilles Deleuze, *Foucault*, trans. Sean Hand (Minneapolis: University of Minnesota Press, 1988), 34.

4. Michel Foucault, "Why Study Power: The Question of the Subject," in *Beyond Structuralism and Hermeneutics*, ed. Hubert L. Dreyfus and Paul Rabinow (Chicago: University of Chicago Press, 1982), 208.

5. See Rosalind Krauss, "Grids," *October* 9 (1979), 50–64.

6. Leon Battista Alberti, *On Painting: A New Translation and Critical Edition*, trans. Rocco Sinisgalli (Cambridge: Cambridge University Press, 2011), 51.

7. See Karin Leonhard and Robert Felfe, *Lochmuster und Linienspiel: Überlegungen zur Druckgrafik des 17. Jahrhunderts* (Freiburg im Breisgau and Berlin: Rombach, 2006), 29–36.

8. See Birgit Schneider, *Textiles Prozessieren: Eine Mediengeschichte der Lochkartenweberei* (Zürich and Berlin: Diaphanes, 2007).

9. Alberti, *On Painting*, 23.

10. Ibid., 51. Opinions differ as to who invented the threaded veil. Some art historians favor Filippo Brunelleschi, to whom Alberti dedicated the Italian version of his treatise *De pictura* (*Della Pittura*). Brunelleschi is said to have used a *velo* to draft the panel of the Florence Baptistery with which he proved the "truth" of central perspective; see Volker Hoffmann, "Filippo Brunelleschi: Kuppelbau und Perspektive," in *Saggi in onore di Renato Bonelli: Quaderni dell'istituto di storia dell'architettura*, ed. Corrado Bozzoni, Giovanni Carbonara, and Gabriela Villetti (Rome: Multigrafica Editrice, 1992), 323. It is known that Brunelleschi used squared paper for the topographic registration of Rome's ancient ruins. According to his biographer Antonio Manetti, Brunelleschi "made measured drawings of Roman buildings, using his understanding of standard surveying techniques. . . . The results were recorded 'on offcuts of parchment . . . by means of squared divisions of the sheets, with arabic numerals and characters which Filippo alone understood'"; Martin Kemp, *The Science of Art: Optical Themes in Western Art from Brunelleschi to Seurat* (New Haven: Yale University Press, 1990), 11–12. The astronomer Paolo dal Pozzo Toscanelli, an alleged friend of Brunelleschi, used squared paper featuring Indian numerals to note line and column to record celestial observations. Antonio di Pietro Averlino, known as Filarete, describes in great detail in *Trattato di architettura* (app. 1461–1464) the usefulness of checkered paper for architectural design, especially with regard to the scalability of to-scale plans that enable readers to have a sense of the dimensions of fictitious buildings; see Filarete, *Tractat über die Baukunst nebst seinen*

Büchern von der Zeichenkunst und den Bauten der Medici, ed. Wolfgang von Oettingen (Vienna, 1890; reprinted, Hildesheim and New York: Olms, 1974), 86.

11. Alberti, *On Painting,* 49.

12. Long before Alberti, such painters as Giotto wrestled with the problem of where to place the nimbus in paintings that conjure up the illusion of depth. Where, for instance, is the nimbus of angels depicted in profile? And where is it in rear views? The Scrovegni Chapel in Padua features halos drawn in perspectival foreshortening. For further details see Wolfgang Braunfeld, *Nimbus und Goldgrund: Wege zur Kunstgeschichte, 1949–1975* (Mittenwald: Mäander Kunstverlag, 1979), 12–15.

13. Hubert Damisch, *The Origin of Perspective,* trans. John Goodmann (Cambridge: MIT Press, 2000), xxi.

14. See Brian Rotman, *Signifying Nothing: The Semiotics of Zero* (Stanford: Stanford University Press, 1993).

15. See Oswald Ashton Wentworth Dilke, "The Culmination of Greek Cartography in Ptolemy," in *Cartography in Prehistoric, Ancient and Medieval Europe and the Mediterranean,* ed. John B. Harley and David Woodward (Chicago: University of Chicago Press, 1987), 177–200.

16. See William Boelhower, "Inventing America: A Model of Cartographic Semiosis," *Word and Image: A Journal of Verbal/Visual Enquiry* 4, no. 2 (1988): 482–83.

17. See Joachim Krausse, "Information at a Glance: On the History of the Diagram," *OASE: Tijdschrift voor architectuur* 48 (1998): 3–30.

18. Martin Heidegger, "The Age of the World Picture," in *The Question Concerning Technology and Other Essays,* trans. William Lovitt (New York: Harper and Row, 1977), 115–54.

19. Hubert Damisch, "La grille comme volonté et comme représentation," in Centre de création industrielle and Centre Georges Pompidou, *Cartes et figures de la Terre,* exhibition catalogue (Paris: Centre de création industrielle and Centre Georges Pompidou, 1980), 30.

20. See James M. Houston, "The Foundation of Colonial Towns in Hispanic America," in *Urbanization and Its Problems: Essays in Honour of E. W. Gilbert,* ed. Robert P. Beckinsale and James M. Houston (Oxford: Blackwell, 1968), 352–90.

21. The discussion that follows draws on Oswald Ashton Wentworth Dilke, "Roman Large-Scale Mapping in the Early Empire," in Harley and Woodward, eds., *Cartography in Prehistoric, Ancient and Medieval Europe and the Mediterranean,* 213–14, 212–33., 216–17, 221–24.

22. Dilke, "Roman Large-Scale Mapping," 215.

23. Damisch, "La grille comme volonté et comme représentation," 32.

24. Aristotle, *Politics,* trans. J. A. Sinclair (Harmondsworth: Penguin, 1992), 1267b23 (p. 134).

25. Ibid.

26. Damisch, "La grille comme volonté et comme représentation," 32.

27. Bernabé Cobo, *Historia de la fundación de Lima* (1639). Quoted in Bertram Lee, ed., *Libros de Cabildos de Lima* (Lima: Torres Aguirre, 1935), 2:475.

28. See Bernhard Siegert, *Passagiere und Papiere: Schreibakte auf der Schwelle zwischen Spanien und Amerika* (Munich: Wilhelm Fink Verlag, 2006), 130–41.

29. Joaquin García Icazbalceta, ed., "Carta del Padre Fray Jerónimo de Mendieta al Ilustre Señor Licenciado Joan de Ovando, del Consejo de S.M. en la Santa y General Inquisición y Visitador e su Real Consejo de Indias," in *Nueva Colección de documentos para la historia de México* (Mexico City: Editorial Salvador Chávez Hayhoe, 1886–92), 123.

30. "[P]ara recoger in pueblos formados y poner en asiento los muchos espanoles que andan vagueando por aquella tierra, no on poco perjuicio del pro común della"; quoted in Norman F. Martin, *Los vagabundos en la Nueva Espana, siglo XVI* (Mexico City: Editorial Jus, 1957), 60.

31. See Angel Rama, *The Lettered City*, trans. John Charles Chasteen (Durham, N.C.: Duke University Press, 1996).

32. See George W. Geib, "The Land Ordinance of 1785: A Bicentennial Review," *Indiana Magazine of History*, 81, no. 1 (1985), 1–13.

33. Vernon Carstensen, "Patterns on the American Land," *Publius* 18, no. 4 (1988), 33.

34. Beginning in the early eighteenth century, land development in Connecticut, Massachusetts, and Maine instituted the township as the central vehicle for expansion. Precise guidelines were established, from the number of families per township and the size of plots to the percentage of land allocated to the local school and the number of acres to be covered with green lawn. See Catherine Maumi, *Thomas Jefferson et le projet du Nouveau Monde* (Paris: Éditions de la Villette, 2007), 84, and Geib, "The Land Ordinance of 1785," 6. Matters were handled differently in the South: "Under the southern system, land was purchased under warrants that merely specified acres and general locations. Survey, if any, usually followed the occupation of an attractive location of the appropriate size" (ibid.).

35. United States Continental Congress, "An Ordinance for Ascertaining the Mode of Disposing of Lands in the Western Territory (20 May 1785)," in *Journals of the Continental Congress, 1774–1789*, vol. 28 (1785), ed. John C. Fitzpatrick (Washington: U.S. Government Printing Office, 1933), 375.

36. Ibid., 376.

37. See Simon Stevin, *The Haven-Finding Art, or, The Way to Find any Hauen or Place at Sea by the Latitude and Variation* (London: G.B.R.N. and R.B., 1599).

38. U.S. Continental Congress, "An Ordinance," 376–77.

39. Geib, "The Land Ordinance of 1785," 12.

40. See Hildegard Binder Johnson, *Order upon the Land: The U.S. Rectangular Land Survey and the Upper Mississippi Country* (New York: Oxford University Press, 1976), 44 and 143.

41. U.S. Continental Congress, "An Ordinance," 376–77.

42. Johnson, *Order upon the Land*, 44.

43. Carstensen, "Patterns on the American Land," 31. See also Stefan Kaufmann, *Soziologie der Landschaft* (Wiesbaden: VS Verlag für Sozialwissenschaften, 2005), 172.

44. On the roots of this merging of the various meanings of *speculation* during the Italian Renaissance, see chapter 7.

45. Johnson, *Order upon the Land*, 219.

46. See Kaufmann, *Soziologie der Landschaft*, 185.

47. Lewis Evans, "Notes Accompanying the General Map of the Middle British Colonies in America," 1755, quoted in *The Shape of the World*, ed. Simon Berthon and Andrew Robinson (London: George Philip, 1991), 152.

48. Maumi, *Thomas Jefferson*, 43. Surveyors' chains were an additional tool for measuring the east-west distances. For a detailed analysis of the transfer of the methods of maritime cartography onto the North American territory, as well as of the imagination of the American landscape as an ocean, see Kaufmann, *Soziologie der Landschaft*, 185–207 and 210–13.

49. See Johnson, *Order upon the Land*, preface, n.p.

50. Carstensen, "Patterns on the American Land," 31.

51. Maumi, *Thomas Jefferson*, 12. On the North American rectangular survey see also William D. Pattison, *Beginnings of the American Rectangular Land Survey Systems, 1784–1800* (New York: Arno Press, 1979), and Norman J. Thrower, *Original Survey and Land Subdivision: A Comparative Study of the Form and Effect of Contrasting Cadastral Surveys* (Chicago: Rand McNally, 1966).

52. Ernst Neufert, *Bauordnungslehre: Handbuch für rationelles Bauen nach geregeltem Mass* (Frankfurt/M. and Berlin: Ullstein, 1961), 95.

53. See Alexander Klose, "Vom Raster zur Zelle—die Containerisierung der modernen Architektur," lecture delivered at Bauhaus University, Weimar, January 17, 2006; in English, "From Grid to Box: The Containerization of Modern Architecture," lecture delivered at Goethe Institute, Prague, October 7, 2005, at the workshop "City-Media-Space." Both versions available at http://www.containerwelt.info./ordner_eigene_texte.html.

54. In 1911 Karl Bührer and Adolf Saager cofounded Die Brücke: Internationales Institut zur Organisierung der geistigen Arbeit (The Bridge: International Institute for the Organization of Intellectual Work), with the goal of solving the problems resulting from the internationalization of the sciences by creating global standards.

55. See Ludwig Mies van der Rohe, "Hochhausprojekt für Bahnhof Friedrichstrasse in Berlin (1922)," in *Frühlicht 1920–1922: Eine Folge für die Verwirklichung des neuen Baugedankens*, ed. Bruno Taut (Berlin, Frankfurt/M., and Vienna: Ullstein, 1963), 213.

56. Walter Prigge, "Typologie und Norm: Zum modernen Traum der industriellen Fertigung von Wohnungen," in *Constructing Utopia: Konstruktionen künstlicher Welten*, ed. Annett Zinsmeister (Berlin and Zürich: Diaphanes, 2005), 74–75.

57. Le Corbusier, "A Dwelling at Human Scale," in *Precisions on the Present State of Architecture and City Planning*, trans. Edith Schreiber Aujame (Cambridge: MIT Press, 1991), 90, 91, 95, 97–98, and 100 (emphases in the original). *Translator's note*: As in the quote above, Le Corbusier's *cellule* is sometimes translated into English as "dwelling" or "unit of dwelling."

58. Le Corbusier, "American Prologue," in *Precisions on the Present State of Architecture and City Planning*, 3.

59. For barracks, see Axel Dossmann, Jan Wenzel, and Kai Wenzel, *Architektur auf Zeit: Baracken, Pavillons, Container* (Berlin: B-Books, 2006); for containers, see Alexan-

der Klose, *Das Container-Prinzip: Wie eine Box unser Denken verändert* (Hamburg: Mare, 2009).

7. WHITE SPOTS AND HEARTS OF DARKNESS: DRAFTING, PROJECTING, AND DESIGNING AS CULTURAL TECHNIQUES

1. Wolfgang Kemp, "Disegno: Beiträge zur Geschichte des Begriffs zwischen 1547 und 1607," *Marburger Jahrbuch* 19 (1974): 225. See also Uwe Westfehling, *Zeichnen in der Renaissance: Entwicklung, Techniken, Formen, Themen* (Cologne: DuMont, 1993), 75 sq.

2. As in the title *Entwerfen und Entwurf: Praxis und Theorie des künstlerischen Schaffensprozesses* [Designing and design: Practice and theory of the creative act], ed. Gundel Mattenklott and Friedrich Weltzien (Berlin: Reimer, 2003).

3. See Westfehling, *Zeichnen in der Renaissance*, 77.

4. Gundel Mattenklott and Friedrich Weltzien, "Einleitung," in Mattenklott and Weltzien, eds., *Entwerfen und Entwurf*, 7.

5. All quoted phrases are from Michael Glasmeier, "Ansichten von Zeichnungen," in Mattenklott and Weltzien, eds., *Entwerfen und Entwurf*, 76 and 83.

6. See Karen-edis Barzman, *The Florentine Academy and the Early Modern State: The Discipline of "Disegno"* (Cambridge and New York: Cambridge University Press, 2000).

7. Bruno Latour, "Drawing Things Together," in *Representation in Scientific Practice*, ed. Michael Lynch and Steve Woolgar (Cambridge: MIT Press, 1990), 52.

8. Glasmeier, "Ansichten von Zeichnungen," 77.

9. See Latour, "Drawing Things Together," 28.

10. Giorgio Vasari, *Vasari on Technique: Being the Introduction to the Three Arts of Design, Architecture, Sculpture and Painting, Prefixed to the Lives of the Most Excellent Painters, Sculptors and Architects*, trans. Louisa Maclehose, ed. G. Baldwin Brown (London: J. M. Dent, 1907), 205.

11. Ibid., 206

12. "It is the goal of our characteristics to employ signs devised in such a way that all conclusions which may be drawn immediately emerge from the words or characters themselves." Gottfried Wilhelm Leibniz, *Fragmente zur Logik*, ed. Franz Schmidt (Berlin: Akademie Verlag, 1960), 93. On the *characteristica universalis* and *scienzia generalis* as a machinally operating *ars inveniendi*, see Bernhard Siegert, *Passage des Digitalen: Zeichenpraktiken der neuzeitlichen Wissenschaften 1500–1900* (Berlin: Brinkmann and Bose, 2003), 171–75.

13. See Leon Battista Alberti, *On Painting: A New Translation and Critical Edition*, trans. Rocco Sinisgalli (Cambridge: Cambridge University Press, 2011), 32–33, 40.

14. *The Notebooks of Leonardo da Vinci*, ed. and trans. Edward MacCurdy (New York: Reynal and Hitchcok, 1938), 2:105.

15. Leonardo da Vinci, "Three studies of water swirls emanating from a concave surface," 88 x 101 mm, undated (W, 12666 r,), ink on royal white paper, reproduced in Leonardo da Vinci, *Das Wasserbuch: Schriften und Zeichnungen*, trans. Marianne Schneider (Munich: Schirmer/Mosel, 1996), plate 24.

16. Leonardo da Vinci, *Notebooks*, 2:112.

17. See Frank Fehrenbach, *Licht und Wasser: Zur Dynamik naturphilosophischer Leitbilder im Werk Leonardo da Vincis* (Tübingen: Wasmuth Verlag, 1997), 310–21.

18. See the introduction to this volume, p. 14.

19. See Pierre Bourdieu, "The Kabyle House or the World Reversed," in *Algeria 1960: The Disenchantment of the World: The Sense of Honour: The Kabyle House or the World Reversed: Essays* (Cambridge: Cambridge University Press, 1979), 133–53.

20. See David Woodward, "Medieval Mappaemundi," in *Cartography in Prehistoric, Ancient and Medieval Europe and the Mediterranean*, ed. John B. Harley and David Woodward (Chicago: University of Chicago Press, 1987), 286–370.

21. See, for example, the portrait of Captain James Cook (1776) by Nathaniel Dance (National Maritime Museum, Greenwich, England; image online at http://nzetc.victoria .ac.nz/tm/scholarly/Bea04Cook-fig-Bea04CookP002a.html, accessed 9 April 2014). The exhibition "Cartes et figures de la Terre" dedicated an entire section to pictures and photos of European master subjects pointing their fingers at maps or globes. See Centre de création industrielle and Centre Georges Pompidou, *Cartes et figures de la Terre*, exhibition catalog (Paris: Centre de création industrielle and Centre Georges Pompidou, 1980), 354–57.

22. Joseph Conrad, *Heart of Darkness*, with foreword by A. N. Wilson (London: Hesperus, 2002), 5.

23. Joseph Conrad, "Geography and Some Explorers," in *Last Essays* (London and Toronto: J. M. Dent, 1926), 24.

24. Further see Christopher Gogwilt, *The Invention of the West: Joseph Conrad and the Double-Mapping of Europe and the Empire* (Stanford: Stanford University Press, 1995), 109–10.

25. Martin Heidegger, "The Age of the World Picture," in *The Question Concerning Technology and Other Essays*, trans. William Lovitt (New York: Harper and Row, 1977), 118. Subsequent page references appear in parentheses in the text.

26. *Translator's note:* Heidegger's original reads "Grundriss der Naturvorgänge" for "ground plan of natural events." As already pointed out by Heidegger's translator William Lovitt ("The Age of the World Picture," 118 n. 6), the word *Riss*, derived from the verb *reissen* ("to tear"), denotes both a plan and a (torn) opening.

27. Alberti, *On Painting*, 42.

28. See Michael Wiemers, *Bildform und Werkgenese: Studien zur zeichnerischen Bildvorbereitung in der italienischen Malerei zwischen 1450 und 1490* (Munich and Berlin: Deutscher Kunstverlag, 1996), 21.

29. Alberti, *On Painting*, 48–49.

30. Ibid., 42.

31. See Samuel Y. Edgerton, *The Renaissance Rediscovery of Linear Perspective* (New York: Basic Books, 1975), 97–104 and especially 106–23.

32. Ibid., 98.

33. Ibid., 107–09.

34. See Gottfried Boehm, *Bildnis und Individuum: Über den Ursprung der Porträtmalerei in der italienischen Renaissance* (Munich: Prestel Verlag, 1985), 240.

35. Leonardo da Vinci, quoted in Edgerton, *Renaissance Rediscovery*, 92.

36. Edgerton, *Renaissance Rediscovery*, 90.

37. Hugo Grotius, *The Free Sea*, trans. Richard Hakluyt, ed. David Armitage (Indianapolis: Liberty Fund, 2004), 13–14.

38. Peter Sloterdijk, *In the World Interior of Capital: For a Philosophical Theory of Globalization*, trans. Wieland Hoban (Cambridge: Polity, 2013), 101.

39. Ibid., 77–80.

40. See Kemp, *Disegno*, 225.

41. See Uwe Granzow, *Quadrant, Kompass und Chronometer: Technische Implikationen des euro–asiatischen Seehandels von 1500 bis 1800* (Stuttgart: Steiner, 1986), 296–97.

42. See Mark Monmonier, *Rhumb Lines and Map Wars: A Social History of the Mercator Projection* (Chicago: University of Chicago Press, 2004), 1–16, 57–61. See also Granzow, *Quadrant, Kompass und Chronometer*, 297.

43. The phrase is used by Carmen C. Bambach to describe drawing; see *Drawing and Painting in the Italian Renaissance Workshop: Theory and Practice, 1300–1600* (Cambridge and New York: Cambridge University Press, 1999), 11.

44. Edgerton, *Renaissance Rediscovery*, 113.

45. See chapter 6 in this volume.

46. Edgerton, *Renaissance Rediscovery*, 114.

47. On *sinopia* see Wiemers, *Bildform und Werkgenese*, 31–58.

48. Alberti, *On Painting*, 50–54.

49. See Bambach, *Drawing and Painting in the Italian Renaissance Workshop: Theory and Practice*, 128–33 and 189–94. The stylus-incised transfer grid of Masaccio's *Trinity* or Paolo Uccello's squared preparatory draft for the *Cenotaph of Sir John Hawkwood* represent formal analogies to Alberti's veil.

50. Edgerton, *Renaissance Rediscovery*, 118.

51. Bambach, *Drawing and Painting*, 10.

52. Ibid., 11.

53. *Translator's note*: A reference to Nietzsche's frequently quoted statement that "our writing tools are also working on our thoughts" ("Unser Schreibzeug arbeitet mit an unseren Gedanken").

54. Francis Ames-Lewis, *Drawing in Early Renaissance Italy*, 2nd ed. (New Haven: Yale University Press, 2000), 25.

55. Alberti, *On Painting*, 82 (translation emended).

56. See Philippe Costamagna, "The Formation of Florentine Draftsmanship: Life Studies from Leonardo and Michelangelo to Pontormo and Salviati," *Master Drawings* 43, no. 3 (2005): 276.

57. See Westfehling, *Zeichnen in der Renaissance*, 79. See also Wiemers, *Bildform und Bildgenese*, 185–91, for the function of Ghirlandaio's *modellos*. In this regard the function of drawing in artisan workshops is completely equivalent to that of draft designs, for instance those of English master shipwright Mathew Baker, who around 1586 was the first to design ships on paper (for further discussion see chapter 8 in this volume).

58. Glasmeier, "Ansichten von Zeichnungen," 76.

59. See the reproduction of Juan de la Cosa's world map in Donald Wigal, Historic Maritime Maps Used for Historic Exploration, 1290–1699 (New York: Parkstone Press, 2000), 54.

60. "Real provision a Amerigo Vespucio," Valladolid, 6 August 1508, Archivo General de Indias (Seville), Indiferente, 1961, libro 1, fol. 66r. See also "Título de Piloto Mayor para Amerigo Vespuche," in José Pulido Rubio, El Piloto mayor de la Casa de la Contratación de Sevilla: Pilotos mayores, Catedraticos de Cosmografia y Cosmografos (Seville: Escuela de Estudios Hispano-Americanos, 1950), 462–63; and "Real Cédula al Piloto mayor Amerigo Despuches [!], dandole poder para examinar, empadronar e rexir a los demas pilotos, e declarallos ábiles o ynábiles para su ofycio," Cadiz, 8 August 1508, in Joaquín Francisco Pacheco, Francisco de Cárdenas y Espejo, and Luis Torres de Mendoza, eds., Colección de documentos ineditos relativos al descubrimiento, conquista y organización de las posesiones españolas en América y Occeanía, first series (Madrid: Imprenta de M. Bernaldo de Quirós, 1864–84), 36:253–54.

61. William Boelhower, "Inventing America: A Model of Cartographic Semiosis," Word and Image: A Journal of Verbal/Visual Enquiry 4, no. 2 (1988): 477.

62. See Pulido Rubio, El Piloto mayor, 257.

63. "Real provision a Amerigo Vespucio," fol. 66r; "Título de Piloto Mayor para Amerigo Vespuche," 463; "Real Cédula al Piloto mayor Amerigo Despuches," 254.

64. Edgerton, Renaissance Rediscovery, 95.

65. As on display in Kittler's heroizing account of Alberti in Friedrich Kittler, Unsterbliche (Munich: Fink, 2004), 14.

66. Hans Blumenberg, "'Imitation of Nature': Toward a Prehistory of the Idea of Creative Being," Qui parle 12, no. 1 (2000): 18.

8. WATERLINES: STRIATED AND SMOOTH SPACES AS TECHNIQUES OF SHIP DESIGN

1. Joseph Furttenbach, Architectura navalis (Ulm, 1629; reprinted, Hildesheim and New York: Olms, 1975), p. 2 of preface.

2. Hans Blumenberg, Shipwreck with Spectator: Paradigm of a Metaphor for Existence, trans. Steven Randall (Cambridge: MIT Press, 1997), 8.

3. The Divine Comedy of Dante Alighieri: Inferno, ed. and trans. Robert M. Durling (New York and Oxford: Oxford University Press, 1996), 405.

4. Ibid.

5. For a critical view of this periodization see David McGee, "From Craftsmanship to Draftsmanship: Naval Architecture and the Three Traditions of Early Modern Design," Technology and Culture 40, no. 2 (1999): 209–14.

6. Further see Bruno Latour, "Drawing Things Together," in Michael Lynch and Steve Woolgar, eds., Representation in Scientific Practice (Cambridge: MIT Press, 1990), 19–68.

7. The concept of the "open object" was invented by Gilbert Simondon and developed further against the backdrop of Paul Valéry's objet ambigu by Lorenz Engell and

myself. See Gilbert Simondon, "Technical Mentality," *Parrhesia* 7 (2009): 17–27, and Lorenz Engell and Bernhard Siegert, "Editorial," *Zeitschrift für Medien- und Kulturforschung* 1 (2011): 5–9.

8. See Sergio Bellabarba, "The Ancient Methods of Designing Hulls," *Mariner's Mirror* 79, no. 3 (1993): 274. See also R. C. Anderson, "Italian Naval Architecture about 1445," *Mariner's Mirror* 11, no. 2 (1925): 135–63.

9. See Richard Barker, "'Many may peruse us': Ribbands, Moulds and Models in the Dockyards," *Revista da Universidade de Coimbra* 34 (1987): 543–46.

10. See Mauro Bondioli, "The Art of Designing and Building Venetian Galleys from the Fifteenth to the Sixteenth Century," in *Boats, Ships and Shipyards: Proceedings of the Ninth International Symposium on Boat and Ship Archaeology, Venice 2000*, ed. Carlo Beltrame (Oxford: Oxbow Books, 2003), 226.

11. See Tim Ingold, *Lines: A Brief History* (London and New York: Routledge, 2007), 41.

12. See Filipe Castro, "Rising and Narrowing: Sixteenth-Century Geometric Algorithms Used to Design the Bottom of Ships in Portugal," *International Journal of Nautical Archaeology* 36, no. 1 (2007), 150–51; Stephen Johnston, "Making Mathematical Practice: Gentlemen, Practitioners and Artisans in Elizabethan England," Ph.D. diss., Cambridge 1994, 122; and especially Bellabarba, "Ancient Methods of Designing Hulls," 274–80.

13. Johnston, "Making Mathematical Practice," 122.

14. See Castro, "Rising and Narrowing," 150.

15. See Bellabarba, "Ancient Methods of Designing Hulls," 280–81.

16. See Johnston, "Making Mathematical Practice," 126.

17. McGee, "From Craftsmanship to Draftsmanship," 215.

18. Ulrich Alertz, "The Venetian Merchant Galley and the System of *Partisoni*—Initial Steps towards Modern Ship Design," in Beltrame, *Boats, Ships and Shipyards*, 213.

19. Johnston, "Making Mathematical Practice," 126.

20. See Johnston, "Making Mathematical Practice," 110.

21. See Richard Barker, "Fragments from the Pepysian Library," *Revista da Universidade de Coimbra* 32 (1985): 161–78. The Portuguese Fernando Oliveira had already published an *Ars Nautica* in 1570 containing elements for the depiction and construction of hulls that can also be found in later English treatises on shipbuilding. See Eric Rieth, "Remarques sur une série d'illustrations de l'*Ars Nautica* (1570) de Fernando Oliveira," *Neptunia* 169 (1988): 36–43.

22. Johnston, "Making Mathematical Practice," 110.

23. See ibid., 111.

24. On Vasari and the *velum*, see chapters 6 and 7.

25. Johnston, "Making Mathematical Practice," 127.

26. Ibid., 129.

27. Pepys Library, PL 2878, S. 493 (seventeenth-century copy), quoted in Johnston, "Making Mathematical Practice," 135.

28. *The Autobiography of Phineas Pett*, ed. W. G. Perrin, Publications of the Navy Records Society, vol. 51 (London: Ballantyne Press, 1918), 95. Pett notes that he was ordered to Hampton Court again in late 1625 to show "plats" of smaller vessels to the King and

Commissioners of the Navy—"to very little purpose and my great trouble and charge" (136).

29. See Peter Jeffrey Booker, *A History of Engineering Drawing* (London: Chatto and Windus, 1963), 69.

30. See James Dodds and James Moore, *Building the Wooden Fighting Ship* (London: Chatham Publishing, 2005), 51.

31. See Anthony Deane, *Deane's Doctrine of Naval Architecture*, ed. Brian Lavery (1670; London: Conway Maritime Press, 1981).

32. Pett, *Autobiography*, 136.

33. See George Kish, "Early Thematic Mapping: The Work of Phillippe Buache," *Imago Mundi* 28 (1976), 129–32.

34. See Josef W. Konvitz, *Cartography in France, 1660–1848: Science, Engineering, and Statecraft* (Chicago: University of Chicago Press, 1987), 75.

35. See Konvitz, *Cartography in France*, 77.

36. François de Dainville, "From the Depths to the Heights: Concerning the Marine Origins of the Cartographic Expression of Terrestrial Relief by Numbers and Contour Lines," *Surveying and Mapping* 30 (1970), 395.

37. Booker, *History of Engineering Drawing*, 71.

38. See Dainville, "From the Depths to the Heights," 398.

39. See Konvitz, *Cartography in France*, 78.

40. Henri Louis Duhamel du Monceau, *Élémens de l'architecture navale, ou Traité pratique de la construction des vaisseaux* (Paris: Charles-Antoine Jombert, 1758), 262.

41. See Alain Corbin, *The Lure of the Sea: The Discovery of the Seaside in the Western World, 1750–1840*, trans. Jocelyn Phelps (Harmondsworth: Penguin, 1994), 2–3.

42. See Larrie D. Ferreiro, *Ships and Science: The Birth of Naval Architecture in the Scientific Revolution, 1600–1800* (Cambridge: MIT Press, 2007), 151.

43. See Fredrik Henrik af Chapman, "Treatise on Shipbuilding," trans. James Inman, in Fredrik Henrik af Chapman, *Architectura navalis mercatoria: The Classic of Eighteenth-Century Naval Architecture* (Mineola, N.Y.: Dover Publications, 2006), 128–52.

44. On Leonardo's experiments with geometric bodies submerged in flowing water, see chapter 7 in this volume.

45. Paul Valéry, "Eupalinos, or, The Architect," in *The Collected Works of Paul Valéry*, vol. 4, ed. Jackson Matthews (New York: Pantheon, 1956), 137–38.

46. Frederik Henrik af Chapman, "Treatise on Shipbuilding", 130.

47. See Ferreiro, *Ships and Science*, 161–63.

48. Norman Bel Geddes, *Horizons* (Boston: Little, Brown, 1932), 38.

49. Ibid., 42.

50. Natasha Pulitzer Los, "The Culture of Ship Design," *Rassegna: Themes in Architecture* 12, no. 44/4 (1990): 9.

51. Quoted in Joachim Krausse and Claude Lichtenstein, "Earthwalking—Skyriding: An Invitation to Join R. Buckminster Fuller on a Voyage of Discovery," in R. Buckminster Fuller et al., *Your Private Sky: R. Buckminster Fuller: Discourse*, ed. Joachim Krausse and Claude Lichtenstein (Baden and Zürich: Springer, 2001), 11.

52. See Carl Schmitt, *The Nomos of the Earth in the International Law of the Jus Publicum Europaeum*, trans. G. L. Ulmen (New York: Telos Press, 2003).

53. "The house is gone." Theodor W. Adorno, *Minima Moralia: Reflections on a Damaged Life*, trans. E. F. N. Jephcott (London: Verso, 2005), 39.

9. FIGURES OF SELF-REFERENCE: A MEDIA GENEALOGY OF THE TROMPE-L'OEIL IN SEVENTEENTH-CENTURY DUTCH STILL LIFE

1. See Martin Heidegger, *Being and Time*, trans. John Macquarie and Edward Robinson (London: SCM Press 1962), 97–107.

2. For example, Pierre Lebrun's statement, "Pour parler des riches peintures, il en faut parler comme si les choses estoient vrayes, non pas peintes." Pierre Lebrun, "Recueil des essaies des merveilles de la peinture (1635)," in Mary Philadelphia Merrifield, ed., *Medieval and Renaissance Treatises on the Arts of Painting: Original Texts with English Translations* (London: John Murray, 1849; reprinted Mineola, N.Y.: Dover Publications, 1999), 825.

3. Cornelis de Bie, *Het gulden Cabinet van de edel vry schilder-const ontsloten door den lanck ghewenschten vrede tusschen de twee machtighe croonen van Spaignien en Vrnnckryck*, 2nd ed. (Antwerp, 1662), 216–17; quoted in Eberhard König and Christiane Schön, eds., *Stilleben* (Berlin: Dietrich Reimer Verlag, 1996), 114–15.

4. Cornelius Norbertus Gijsbrecht, *Reverse Side of a Painting*. Oil on canvass, 66.5 x 86.5 cm, Kopenhagen, Statens Museum for Kunst.

5. Louis Marin, "Representation and Simulacrum," in *On Representation*, trans. Catherine Porter (Stanford: Stanford University Press, 2001), 316 (emphasis in the original).

6. See Jean Baudrillard, "The Trompe-l'Oeil," in *Calligram: Essays in New Art History from France*, ed. Norman Bryson (Cambridge: Cambridge University Press, 1988), 53.

7. Svetlana Alpers, *The Art of Describing: Dutch Art in the Seventeenth Century* (Chicago: University of Chicago Press, 1983), 90.

8. See Jochen Sander, ed., *Die Magie der Dinge: Stillebenmalerei 1500–1800*, exhibition catalog of the Städel Museum, Frankfurt a. M., and Kunstmuseum Basel (Ostfildern, Germany: Hatje Cantz, 2008), 64.

9. Lee Hendrix and Thea Vignau-Wilberg, eds., *Mira calligraphiae monumenta: A Sixteenth-Century Calligraphic Manuscript Inscribed by Georg Bocskay and Illuminated by Joris Hoefnagel* (Los Angeles: J. Paul Getty Museum, 1992), fol. 37r.

10. Friedrich Teja Bach has discovered that Albrecht Dürer's drawings in the margins of Maximilian I's prayer book use the transparency of recto and verso for metamorphic correspondences. However, unlike Hoefnagel, Dürer was more concerned with creating allegories than three-dimensional illusions. Folio 51 of the prayer book is of special interest in the context of Hoefnagel's flower trompe-l'oeils: while the recto depicts Mary, the verso features an allegorical representation of wisdom as an old female farmer. While the women differ substantially, the stems of the flower consoles they are standing on are identical. See Friedrich Teja Bach, *Struktur und Erscheinung: Untersuchungen zu Dürers graphischer Kunst* (Berlin: Gebr. Mann Verlag, 1996), 194–203, esp. 194.

11. Referring to Dürer's "recto-verso-paraphrases," Bach draws attention to the reality effect that emerges from spatial juxtaposition: "By literally contrasting Maria and the figure of wisdom on the back and front of one and the same transparent vellum page, the knowledge of the typological correspondence is transformed into a real picture." See Bach, *Struktur und Erscheinung*, 200.

12. See Ingvar Bergström, *Dutch Still Life Painting in the Seventeenth Century*, trans. Christina Hedström and Gerald Taylor (London: Faber and Faber, 1956), 8; Hulin de Loo, "La vignette chez les enlumineurs gantois entre 1470 et 1500," *Bulletin de la classe de Beaux Arts de L'Academie Royale de Belgique* 21 (1939): 158–80; Otto Pächt, *The Master of Mary of Burgundy* (London: Faber and Faber, 1948); and Jonathan J. G. Alexander, *The Master of Burgundy: A Book of Hours for Engelbert of Nassau* (Oxford: Bodleian Library, 1970). For more recent appraisals see Frank O. Büttner, "Ikonographisches Eigengut der Randzier in spätmittelalterlichen Handschriften: Inhalte und Programme," *Skriptorium* 39 (1985): 197–233; Thomas DaCosta Kaufmann and Virginia Roehrig Kaufmann, "The Sanctification of Nature: Observations on the Origins of Trompe l'Oeil in Netherlandish Book Painting of the Fifteenth and Sixteenth Centuries," *J. Paul Getty Museum Journal* 19 (1991): 43–64; Michael Camille, *Image on the Edge: The Margins of Medieval Art* (London: Reaktion Books, 1992); Myra D. Orth, "What Goes Around: Borders and Frames in French Manuscripts," in "Essays in Honor of Lilian M. C. Randall," special issue, *Journal of the Walters Art Gallery* 54 (1996):189–203; Bodo Brinkmann, *Die flämische Buchmalerei am Ende des Burgunderreichs: Der Meister des Dresdener Gebetbuchs und die Miniaturisten seiner Zeit* (Turnhout, Belgium: Brepols, 1997), 2 vols.; Ann Margreet As-Vijvers, "More than Marginal Meaning? The Interpretation of Ghent-Bruges Border Decoration," *Oud Holland* 116 (2003): 3–33 (with many thanks to Helga Lutz).

13. See Dagmar Thoss, "Georg Hoefnagel und seine Beziehungen zur Gent-Brügger Buchmalerei," *Jahrbuch der Kunsthistorischen Sammlungen in Wien* 82/83 (1986/87): 199–211. DaCosta Kaufmann and Roehrig Kaufmann also view Hoefnagel as the "penultimate stage" in the development of independent still-life painting. See DaCosta Kaufmann and Roehrig Kaufmann, "The Sanctification of Nature," 46–47.

14. Pächt, *Master of Mary of Burgundy*, 25.

15. Ibid.

16. See Thomas Kren, "Revolution and Transformation: Painting in Devotional Manuscripts, circa 1467–1485," in Thomas Kren and Scot McKendrick, eds., *Illuminating the Renaissance: The Triumph of Flemish Manuscript Painting in Europe* (Los Angeles: J. Paul Getty Museum, 2003), 126.

17. Thoss, "Georg Hoefnagel," 201.

18. DaCosta Kaufmann and Roehrig Kaufmann trace the illusionary borders of the Ghent-Bruges style back to the devotional practices associated with hour books. It appears that many of the latter were taken along on pilgrimages and served as a collection and storage medium for various objects, especially for pilgrims' badges and small pictures, as well as for pressed plants endowed with symbolic meaning. As revealed by puncture marks in surviving hour books, badges and pictures were sown into or attached to the pages. The Kaufmanns argue that the trompe-l'oeils on the margins of

the Ghent-Bruges manuscripts imitate these devotionals (see DaCosta Kaufmann and Roehrig Kaufmann, "The Sanctification of Nature," 57). As persuasive as it may sound, the thesis falls short of the mimesis theory under discussion here. The imitation of real objects is restricted to those that happen to be flat, such as pictures, emblems, and badges. Yet DaCosta Kaufmann and Roehrig Kaufmann also claim that a page featuring Saint Jacob's mussels imitates empirical practices. However, there is no evidence for this. There are only painted holes for Saint Jacob's mussels, which comes as no surprise given that mussels are rounded (i.e., non-flat) three-dimensional objects liable to destroy the spine of a book one wishes to close. The authors tend to elide the difference between two- and three-dimensional objects that is essential to the trompe-l'oeil effect. There is a categorical difference between pressed and nonpressed objects. A trompe-l'oeil is not just the imitation of a real object; it is not a matter of simple mimesis, but an instance of excessive mimesis that transcends the devotional practices associated with hour books.

19. As a vellum manuscript border motive, dragonflies can be traced back to the fourteenth century, though not as trompe-l'oeils. Cf. the so-called Luttrell Psalter, made for Geoffrey Luttrell (East Anglia, c. 1340; British Museum, London). The bottom border of folio 36v shows a hybrid man sticking out his tongue at a large dragonfly. See Lilian M. C. Randall, *Images in the Margins of Gothic Manuscripts* (Berkeley and Los Angeles: University of California Press, 1966), figure 349.

20. James H. Marrow, "Scholarship on Flemish Manuscript Illumination of the Renaissance: Remarks on Past, Present, and Future," in Elizabeth Morrison and Thomas Kren, eds., *Flemish Manuscript Painting in Context: Recent Research* (Los Angeles: Getty Publications, 2006), 163–76, esp. 170.

21. DaCosta Kaufmann and Roehrig Kaufmann, "The Sanctification of Nature," 61.

22. See Maurits Smeyers, *Flemish Miniatures from the Eighth to the Mid-Sixteenth Century* (Leuven, Belgium: Davidsfonds, 1999), 420.

23. Pächt, *Master of Mary of Burgundy*, 33.

24. Wiener Codex 1857. Bodo Brinkmann speculates that the owner of the Vienna Codex was Margaret of York; see his *Die flämische Buchmalerei am Ende des Burgunderreiches*, 23–24.

25. Manfred Schneider, "Luther with McLuhan," in *Religion and Media*, ed. Hent de Vries and Samuel Weber (Stanford: Stanford University Press, 2001), 210.

26. See also Thoss, "Georg Hoefnagel," 202: "The basic idea is that in principle the border boasts a higher degree of reality than the scene depicted in the centre."

27. "[T]he tiny spark of contingency, of the here and now, with which reality . . . seared the subject." Walter Benjamin, "Little History of Photography," trans. M. W. Jephcott and K. Shorter, in *Walter Benjamin: Selected Writings*, ed. M. W. Jennings, H. Eiland, and G. Smith (Cambridge: Harvard University Press, Belknap Press, 1999), 2:510.

28. Victor Stoichita, *The Self-Aware Image: An Insight into Early Modern Meta-Painting* (Cambridge: Cambridge University Press, 1997), 23 (emphases in the original).

29. Jean-François Lyotard, *Discourse, Figure*, trans. Antony Hudek and Mary Lydon (Minneapolis: University of Minnesota Press, 2011), 192.

30. See ibid., 163.

31. See Emmanuel Alloa, "Jean-François Lyotard: Der Durchbruch des Mediums," *Artnet Magazine*, 22 May 2007, http://www.artnet.de/magazine/features/alloa/alloa05-22-07.asp (accessed 7 November 2009).

10. DOOR LOGIC, OR, THE MATERIALITY OF THE SYMBOLIC: FROM CULTURAL TECHNIQUES TO CYBERNETIC MACHINES

1. Theodor W. Adorno, *Minima Moralia: Reflections on a Damaged Life*, trans. E. F. N. Jephcott (London: Verso, 1974, 2005), 40.

2. Jacques Lacan, *The Seminar of Jacques Lacan, Book 2: The Ego in Freud's Theory and in the Technique of Psychoanalysis, 1954–1955*, trans. Sylvana Tomaselli, with notes by John Forrester (Cambridge: Cambridge University Press, 1988), 47. See also Friedrich Kittler, "The World of the Symbolic—A World of the Machine," *Literature, Media, Information Systems: Essays*, ed. John Johnston, trans. Stephanie Harris (Amsterdam: OAP, 1997), 130–46.

3. *Translator's note*: A more specific term than its English cognate *gate*, German *Gatter* primarily refers to a gate that is part of a fence, fold, or corral. As Siegert will point out toward the end of this chapter, the term now also refers to an electronic gate as part of a circuit or switchboard.

4. For example, in the cave at Pech Merle in southwestern France there is a representation of the step-by-step metamorphosis of a bison into a woman. See André Leroi-Gourhan, *Treasures of Prehistoric Art*, trans. Norbert Guterman (New York: Abrams, 1967), 420, images 368–71. See also p. 521.

5. Gottfried Semper, *Style in the Technical and Tectonic Arts; or, Practical Aesthetics,* trans. Harry Francis Mallgrave and Michael Robinson (Los Angeles: Getty Publications, 2004), 247 (emphasis in the original).

6. See Dirk Baecker, "Die Dekonstruktion der Schachtel: Innen und Aussen in der Architektur," in *Unbeobachtbare Welt: Über Kunst und Architektur*, ed. Niklas Luhmann, Frederick D. Bunsen, and Dirk Baecker (Bielefeld: Haux, 1990), 83.

7. "Precisely because it can also be opened, its closure provides the feeling of a stronger isolation against everything outside this space than the mere unstructured wall. The latter is mute, but the door speaks." Georg Simmel, "Bridge and Door," trans. Mark Rotter, *Theory, Culture and Society* 11 (1994): 7.)

8. Baecker, "Die Dekonstruktion der Schachtel," 91.

9. Franz Kafka, "Before the Law," in *"The Metamorphosis," "In the Penal Colony," and Other Stories*, trans. Joachim Neugroschel (New York: Scribner, 1995), 148–49.

10. See Wolfgang Schäffner, "Punto, línea, abertura: Elementos para una historia medial de la arquitectura y el diseño," Walter Gropius Chair Inaugural Lecture Series, 2003/2004, FADU, University of Buenos Aires; typescript in possession of Bernhard Siegert.

11. Lacan, *The Seminar, Book 2: The Ego in Freud's Theory*, 302.

12. Cicero, *De natura deorum*, 2:27, quoted in "Thyra," in *Paulys Realencyclopädie der classischen Altertumswissenschaften: Neue Bearbeitung*, ed. Georg Wissowa, Wilhelm Kroll, and Karl Mittelhaus, 2nd series, Halbband 11 (Stuttgart: J. B. Metzler, 1936), 740.

13. John Arnott MacCulloch, "Door," *Encyclopaedia of Religion and Ethics*, ed. James Hastings (New York: Scribners, 1912), 4:846.

14. Arnold van Gennep, *The Rites of Passage*, trans. Monika Vizedom and Gabrielle L. Caffee (London: Routledge, 1977), 20.

15. Further see Kurt Klusemann, *Das Bauopfer: Eine ethnographisch-prähistorisch-linguistische Studie* (Graz-Hamburg: Selbstverlag, 1919).

16. ". . . terrible instrument qu'on ne doit manier qu'à bon escient et selon les rites et qui'il faut entourer de toutes les garanties magiques"; Marcel Griaule, "Seuil," in *Documents: Archéologie, Beaux-Arts, Ethnographie, Variétés* 2, no. 2 (1930), 103.

17. The left wing is not the work of Campin but was presumably created by Campin's pupil Rogier van der Weyden and replaced the original left wing (which is lost) after the donor Peter Engelbrecht had married for the second time. See Felix Thürlemann, *Robert Campin, das Merode-Triptychon: Ein Hochzeitsbild für Peter Engelbrecht und Gretchen Schrinmechers aus Köln* (Frankfurt/M.: Fischer Taschenbuch Verlag, 1997), 10 sq.

18. See Helga Lutz, "Medien des Entbergens: Falt- und Klappoperationen in der altniederländischen Kunst des späten 14. und frühen 15. Jahrhunderts," in *Archiv für Mediengeschichte* 10 (2010): 32.

19. See Georges Didi-Huberman, *Phasmes: Essais sur l'apparition* (Paris: Éditions de Minuit, 1998), 9.

20. Susie Nash, *Northern Renaissance Art* (Oxford and New York: Oxford University Press, 2008), 246.

21. Lutz, "Medien des Entbergens," 32.

22. Karl Schade, *Ad excitandum devotionis affectum: Kleine Triptychen in der altniederländischen Malerei* (Weimar: VDG, 2001), 96.

23. See Lutz, "Medien des Entbergens," 32.

24. Compare Samuel van Hoogstraten's *View into a Corridor*, where the flight of thresholds extends into a painting on the back wall that shows an open four-poster bed.

25. Robert Musil, "Doors and Portals," in *Posthumous Papers of a Living Author*, trans. Peter Wortsman (Hygiene, Colo.: Eridanos, 1987), 59.

26. James Clerk Maxwell, letter to Peter Guthrie Tait, 11 December 1867, in *The Scientific Letters and Papers of James Clerk Maxwell*, ed. P. M. Harman (Cambridge: Cambridge University Press, 1995), 2:331.

27. Musil, "Doors and Portals," 57.

28. Ibid., 59.

29. See James Buzard, "Drehtür: Permanente Umwälzungen," *ARCH+: Zeitschrift für Architektur und Städtebau*, no. 191/192 (March 2009), 40.

30. Quoted in James Buzard, "Perpetual Revolution," *Modernism/Modernity* 8, no. 4 (2001): 561.

31. Musil, "Doors and Portals," 59.

32. Adorno, *Minima Moralia*, 42.

33. Ibid.

34. Achim Pietzcker, "Schiebetür," *Arch+: Zeitschrift für Architektur und Städtebau*, no. 191/192 (March 2009), 91.

35. Ibid.

36. Lacan, *The Seminar, Book 2: The Ego in Freud's Theory*, 302.

37. Ibid.

38. Jacques Lacan, *The Seminar of Jacques Lacan, Book 3: The Psychoses, 1955–1956*, ed. Jacques-Alain Miller, trans. Russell Grigg (New York: Norton, 1997), 150.

39. David Cronenberg, dir., *Videodrome* (Canadian Film Development Corporation, Montreal and Toronto, 1983).

40. Roman Jakobson, "Shifters, Verbal Categories, and the Russian Verb," in *Selected Writings*, vol. 2: *World and Language* (The Hague: Mouton, 1971), 133. See also Manfred Riepe, *Bildgeschwüre: Körper und Fremdkörper im Kino David Cronenbergs: Psychoanalytische Filmlektüren nach Freud und Lacan* (Bielefeld: transcript, 2002), 87–119, esp. 109.

41. "In its nature, the door belongs to the symbolic order, and it opens up either on to the real, or the imaginary, we don't know quite which, but it is either one or the other." Lacan, *The Seminar, Book 2: The Ego in Freud's Theory*, 302.

BIBLIOGRAPHY

Adorno, Theodor W. *Minima Moralia: Reflections on a Damaged Life*. Translated by E. F. N. Jephcott. London: Verso, 2005.

Alberti, Leon Battista. *On Painting: A New Translation and Critical Edition*. Translated and with an introduction by Rocco Sinisgalli. Cambridge: Cambridge University Press, 2011.

Alertz, Ulrich. "The Venetian Merchant Galley and the System of *Partisoni*—Initial Steps towards Modern Ship Design." In *Boats, Ships and Shipyards: Proceedings of the Ninth International Symposium on Boat and Ship Archaeology, Venice 2000*, edited by Carlo Beltrame, 212–21. Oxford: Oxbow Books, 2003.

Alexander, Jonathan J. G. *The Master of Burgundy: A Book of Hours for Engelbert of Nassau*. Oxford: Bodleian Library, 1970.

Alliès, Paul. *L'invention du territoire*. Grenoble: Presses Universitaires de Grenoble, 1960.

Alloa, Emmanuel. "Jean-François Lyotard: Der Durchbruch des Mediums." *Artnet Magazine*, 22 May 2007. http://www.artnet.de/magazine/features/alloa/alloa05-22-07.asp (accessed 7 November 2009).

Alpers, Svetlana. *The Art of Describing: Dutch Art in the Seventeenth Century*. Chicago: University of Chicago Press, 1983.

Ames-Lewis, Francis. *Drawing in Early Renaissance Italy*. 2nd ed. New Haven: Yale University Press, 2000.

Anderson, R. C. "Italian Naval Architecture about 1445." *Mariner's Mirror* 11, no. 2 (1925): 135–63.

Apollonius Rhodos. *Argonautica*. Translated by William H. Race. Cambridge: Harvard University Press, 2005.

Aristotle. *Historia Animalium*. Translated by D'Arcy Wentworth Thompson. Vol. 4 of *The Works of Aristotle*, edited by J. A. Smith and W. D. Ross. Oxford: Clarendon Press, 1910.

———. *The Politics* [1253a7-10]. Translated by J. A. Sinclair. Harmondsworth: Penguin, 1992.

As-Vijvers, Ann Margreet. "More than Marginal Meaning? The Interpretation of Ghent-Bruges Border Decoration." *Oud Holland* 116 (2003): 3–33.

Augustus. "Res gestae divi Augusti." In Sextus Aurelius Victor, *De vita et moribus imperatorum romanorum: Excerpta ex libris Sexti Aurelii Victoris, à Caesare Augusto usque ad Theodosium imperatorem*, edited by Andreas Schott. Antwerp: Ex officina Christophori Plantini, 1579.

Bach, Friedrich Teja. *Struktur und Erscheinung: Untersuchungen zu Dürers graphischer Kunst.* Berlin: Gebr. Mann Verlag, 1996.

Baecker, Dirk. "Die Dekonstruktion der Schachtel: Innen und Aussen in der Architektur." In *Unbeobachtbare Welt: Über Kunst und Architektur*, edited by Niklas Luhmann, Frederick D. Bunsen, and Dirk Baecker, 67–104. Bielefeld, Germany: Haux, 1990.

Bambach, Carmen C. *Drawing and Painting in the Italian Renaissance Workshop: Theory and Practice, 1300–1600.* Cambridge and New York: Cambridge University Press, 1999.

Barker, Richard. "Fragments from the Pepysian Library." *Revista da Universidade de Coimbra* 32 (1985): 161–78.

———. "'Many May Peruse Us': Ribbands, Moulds and Models in the Dockyards." *Revista da Universidade de Coimbra* 34 (1987): 539–59.

Barnes, Julian. *Flaubert's Parrot.* New York: Alfred A. Knopf, 1985.

Barzman, Karen-edis. *The Florentine Academy and the Early Modern State: The Discipline of "Disegno."* Cambridge and New York: Cambridge University Press, 2000.

Baudrillard, Jean. "The Trompe-l'Oeil." In *Calligram: Essays in New Art History from France*, edited by Norman Bryson, 27–52. Cambridge and New York: Cambridge University Press, 1988.

Bel Geddes, Norman. *Horizons.* Boston: Little, Brown, 1932.

Bellabarba, Sergio. "The Ancient Methods of Designing Hulls." *Mariner's Mirror* 79, no. 3 (1993): 274–92.

Belting, Hans. *Bild–Anthropologie: Entwürfe für eine Bildwissenschaft.* Munich: Wilhelm Fink Verlag, 2001.

Benjamin, Walter. "Little History of Photography." Translated by M. W. Jephcott and K. Shorter. In *Walter Benjamin: Selected Writings*, edited by M. W. Jennings, H. Eiland, and G. Smith, 2:507–30. Cambridge: Harvard University Press, Belknap Press, 1999.

Benoit, Fernand. "Jas d'ancre à tête de Meduse." *Revue archéologique* 6, no. 37 (1951): 223–28.

Bense, Max. *Einführung in die informationstheoretische Ästhetik: Grundlagen und Anwendungen in der Texttheorie.* Reinbek: Rowohlt, 1969.

———. "Jabberwocky: Text und Theorie, Folgerungen zu einem Gedicht von Lewis Carroll." In *Radiotexte: Essays, Vorträge, Hörspiele*, edited by Caroline Walther and Elisabeth Walther, 63–83. Heidelberg: Winter, 2000.

Bense, Max, and Ludwig Harig. "Der Monolog der Terry Jo." In *Neues Hörspiel: Texte, Partituren*, edited by Klaus Schöning. Frankfurt/M.: Suhrkamp, 1969.

Bergström, Ingvar. *Dutch Still-Life Painting in the Seventeenth Century.* Translated by Christina Hedström and Gerald Taylor. London: Faber and Faber, 1956.

Berthon, Simon, and Andrew Robinson, eds. *The Shape of the World.* London: George Philip, 1991.

Blumenberg, Hans. *Shipwreck with Spectator: Paradigm of a Metaphor for Existence.* Translated by Steven Rendall. Cambridge: MIT Press, 1997.

———. "'Imitation of Nature': Toward a Prehistory of the Idea of Creative Being." *Qui Parle* 12, no. 1 (2000): 17–54.

Boehm, Gottfried. *Bildnis und Individuum: Über den Ursprung der Porträtmalerei in der italienischen Renaissance.* Munich: Prestel Verlag, 1985.

Böhme, Gernot, and Hartmut Böhme. *Feuer, Wasser, Erde, Luft.* Munich: Beck, 1996.

Böhme, Hartmut, Peter Matussek, and Lothar Müller. *Orientierung Kulturwissenschaft: Was sie kann, was sie will.* rororo series, 55608. Reinbek: Rowohlt, 2000.

Boelhower, William. "Inventing America: A Model of Cartographic Semiosis." *Word and Image: A Journal of Verbal/Visual Enquiry* 4, no. 2 (1988): 475–97.

Bolkestein, Hendrik, and Adolph Kalsbach. "Armut I." In Klausman et al., *Reallexikon für Antike und Christentum*, vol. 1, columns 698–705.

Bondioli, Mauro. "The Art of Designing and Building Venetian Galleys from the Fifteenth to the Sixteenth Century." In *Boats, Ships and Shipyards: Proceedings of the Ninth International Symposium on Boat and Ship Archaeology, Venice 2000*, edited by Carlo Beltrame, 222–27. Oxford: Oxbow Books, 2003.

Booker, Peter Jeffrey. *A History of Engineering Drawing.* London: Chatto and Windus, 1963.

Bosse, Heinrich. "Dichter kann man nicht bilden: Zur Veränderung der Schulrhetorik nach 1770." *Jahrbuch für Internationale Germanistik* 10 (1978): 80–125.

———. "'Die Schüler müssen selbst schreiben lernen,' oder, Die Einrichtung der Schiefertafel." In *Schreiben—Schreiben lernen: Rolf Sanner zum 65. Geburtstag*, edited by Dietrich Boueke and Norbert Hopster, 164–99. Tübinger Beiträge zur Linguistik 249. Tübingen: Narr, 1985.

Bourdieu, Pierre. "The Kabyle House or the World Reversed." In *Algeria 1960: The Disenchantment of the World: The Sense of Honour: The Kabyle House or the World Reversed: Essays*, translated by Richard Teese, 133–53. Cambridge and New York: Cambridge University Press, 1979.

Braunfeld, Wolfgang. *Nimbus und Goldgrund: Wege zur Kunstgeschichte, 1949–1975.* Mittenwald, Germany: Mäander Kunstverlag, 1979.

Brentjes, Burchard. *Die Erfindung des Haustieres.* 3rd ed. Leipzig: Urania-Verlag, 1986.

Brinkmann, Bodo. *Die flämische Buchmalerei am Ende des Burgunderreichs: Der Meister des Dresdener Gebetbuchs und die Miniaturisten seiner Zeit.* 2 vols. Turnhout, Belgium: Brepols, 1997.

Brunus, Conradus [Konrad Braun]. "De legationibus libri quinque: Cunctis in repub. versantibus, aut quolibet magistratu fungentibus perutiles, & lectu iucundi." In *D. Conradi Bruni iureconsulti opera tria.* Mainz: Francis Behem, 1548.

Büttner, Frank O. "Ikonographisches Eigengut der Randzier in spätmittelalterlichen Handschriften: Inhalte und Programme." *Skriptorium* 39 (1985): 197–233.

Busbecq, Ogier Ghiselin de. *The Turkish Letters of Ogier Ghiselin de Busbecq, Imperial Ambassador at Constantinople, 1554–1562.* Translated by Edward Seymour Forster. Oxford: Clarendon Press, 1927; 1968.

Burioni, Matteo. "Gattungen, Medien, Techniken: Vasaris Einführung in die drei Künste des *disegno*." In *Einführung in die Künste der Architektur, Bildhauerei und Malerei* by Giorgio Vasari, edited and translated by Matteo Burioni, 7–24. Berlin: Verlag Klaus Wagenbach, 2006.

Busch, Werner. *Joseph Wright of Derby: Das Experiment mit der Luftpumpe, eine Heilige Allianz zwischen Wissenschaft und Religion*. Frankfurt/M: Fischer Taschenbuch, 1986.

Buzard, James. "Perpetual Revolution." *Modernism/Modernity* 8, no. 4 (2001): 559–81.

———. "Drehtür: Permanente Umwälzungen." *Arch+: Zeitschrift für Architektur und Städtebau*, no. 191/192 (March 2009): 39–44.

Camille, Michael. *Image on the Edge: The Margins of Medieval Art*. London: Reaktion Books, 1992.

Campe, Rüdiger. "Pronto! Telefonate und Telefonstimmen." In *Diskursanalysen I: Medien*, edited by Friedrich A. Kittler et al., 68–93. Opladen: Westdeutscher Verlag, 1987.

Campe, Rüdiger. *Affekt und Ausdruck: Zur Umwandlung der literarischen Rede im 17. und 18. Jahrhundert*. Tübingen: Niemeyer, 1990.

Canetti, Elias. *Crowds and Power*. Translated by Carol Stewart. New York: Viking, 1962.

Carstensen, Vernon. "Patterns on the American Land." *Publius* 18, no. 4 (1988): 31–39.

Cassirer, Ernst. *The Philosophy of Symbolic Forms*. Vol. 1, *Language*, translated by Ralph Manheim. New Haven: Yale University Press, 1955.

Castro, Filipe. "Rising and Narrowing: Sixteenth-Century Geometric Algorithms Used to Design the Bottom of Ships in Portugal." *The International Journal of Nautical Archaeology* 36, no. 1 (2007): 148–54.

Centre de création industrielle and Centre Georges Pompidou. *Cartes et figures de la Terre*. Exhibition catalog. Paris: Centre de création industrielle and Centre Georges Pompidou,1980.

Chapman, Frederik Henrik af. "Treatise on Shipbuilding." Translated by John Inman. In *Architectura Navalis Mercatoria: The Classic of Eighteenth-Century Naval Architecture*, 125–52. Mineola, N.Y.: Dover Publications, 2006; originally published in English as *A Treatise on Shipbuilding, with Explanations Regarding the Architectura Navalis Mercatoria* (Cambridge: Printed by J. Smith, sold by Deighton and Sons, 1820).

Clarke, Bruce. "Constructing the Subjectivity of the Quasi-Object: Serres through Latour." Lecture delivered at "Constructions of the Self: The Poetics of Subjectivity," University of South Carolina, April 1999.

Conrad, Joseph. "Geography and Some Explorers." In *Last Essays*. London and Toronto: J. M. Dent, 1926.

———. *Heart of Darkness*. London and New York: Penguin, 1985.

Corbin, Alain. *The Lure of the Sea: The Discovery of the Seaside in the Western World, 1750–1840*. Translated by Jocelyn Phelps. Harmondsworth: Penguin, 1994.

Costamagna, Philippe. "The Formation of Florentine Draftsmanship: Life Studies from Leonardo and Michelangelo to Pontormo and Salviati." *Master Drawings* 43, no. 3 (2005), 274–91.

Dainville, Francois de. "From the Depths to the Heights: Concerning the Marine Origins of the Cartographic Expression of Terrestrial Relief by Numbers and Contour Lines." *Surveying and Mapping* 30 (1970): 389–403.

Damisch, Hubert. "La grille comme volonté et comme représentation." In Centre de création industrielle and Centre Georges Pompidou, *Cartes et figures de la Terre*, 30–40.

———. *The Origin of Perspective*. Translated by John Goodmann. Cambridge: MIT Press, 2000.

Dante Alighieri. *The Divine Comedy of Dante Alighieri: Inferno*. Edited and translated by Robert M. Durling. New York and Oxford: Oxford University Press, 1996.

———. *De vulgari eloquentia*. Edited and translated by Steven Botterill. Cambridge and New York: Cambridge University Press, 1996.

Deane, Anthony. *Deane's Doctrine of Naval Architecture, 1670*. Edited by Brian Lavery. London: Conway Maritime Press, 1981. Originally published 1670.

Deleuze, Gilles. *Foucault*. Translated by Sean Hand. Minneapolis: University of Minnesota Press, 1988.

Deleuze, Gilles, and Félix Guattari. *A Thousand Plateaus: Capitalism and Schizophrenia*. Translated and with an introduction by Brian Massumi. Minneapolis: University of Minnesota Press, 1987.

Derrida, Jacques. *Of Grammatology*. Translated by Gayatri Chakravorty Spivak. Baltimore: Johns Hopkins University Press, 1976.

———. "Signature Event Context." Translated by Samuel Weber and Jeffrey Mehlman. In *Limited Inc* (Evanston, Ill.: Northwestern University Press, 1977), 1-23.

———. *The Post Card: From Socrates to Freud and Beyond*. Translated by Alan Bass. Chicago: University of Chicago Press, 1987.

———. *Rogues: Two Essays on Reason*. Translated by P.-A. Brault and M. Naas. Stanford: Stanford University Press, 2005.

Descartes, René. *Discourse on Method*. Translated by Richard Kennington. Newburyport, Mass.: R. Pullins, 2007.

Didi-Huberman, Georges. *La peinture incarnée*. Paris: Éditions de Minuit, 1985.

———. *Phasmes: Essais sur l'apparition*. Paris: Éditions de Minuit, 1998.

Dilke, Oswald Ashton Wentworth. "The Culmination of Greek Cartography in Ptolemy." In Harley and Woodward, eds., *Cartography in Prehistoric, Ancient, and Medieval Europe and the Mediterranean*, 177–200.

———. "Roman Large-Scale Mapping in the Early Empire." In Harley and Woodward, eds., *Cartography in Prehistoric, Ancient, and Medieval Europe and the Mediterranean*, 212–33.

Dodds, James, and James Moore. *Building the Wooden Fighting Ship*. London: Chatham Publishing, 2005.

Dossmann, Axel, Jan Wenzel, and Kai Wenzel. *Architektur auf Zeit: Baracken, Pavillons, Container*. Berlin: B-Books, 2006.

Duhamel du Monceau, Henri-Louis. *Élémens de l'architecture navale, ou Traité pratique de la construction des vaisseaux*. Paris: Charles-Antoine Jombert, 1758.

Eco, Umberto. "Function and Sign: The Semiotics of Architecture." In *Rethinking Architecture: A Reader in Cultural Theory*, edited by Neil Leach, 173–195. London: Routledge, 1997.

Edgerton, Samuel Y. *The Renaissance Rediscovery of Linear Perspective*. New York: Basic Books, 1975.

Elias, Norbert. *The History of Manners*. Translated by Edmund Jephcott. Vol. 1 of *The Civilizing Process*. New York: Urizen Books, 1978.

Engell, Lorenz, and Bernhard Siegert. "Editorial." *Zeitschrift für Medien- und Kulturforschung*, no. 1 (2011): 5–9.

Fehrenbach, Frank. *Licht und Wasser: Zur Dynamik naturphilosophischer Leitbilder im Werk Leonardo da Vincis*. Tübingen: Wasmuth Verlag, 1997.

Ferreiro, Larrie D. *Ships and Science: The Birth of Naval Architecture in the Scientific Revolution, 1600–1800*. Cambridge: MIT Press, 2007.

Filarete [Antonio di Pietro Averlino]. *Tractat über die Baukunst nebst seinen Büchern von der Zeichenkunst und den Bauten der Medici*. Edited by Wolfgang von Oettingen. Quellenschriften für Kunstgeschichte und Kunsttechnik des Mittelalters und der Neuzeit, new series, 3. Vienna, 1890; reprinted, Hildesheim and New York: Olms, 1974.

Flaubert, Gustave. *Correspondance*. 9 vols. Paris: Louis Conard, 1926–1933.

———. *The Letters of Gustave Flaubert, 1847–1852*. Edited and translated by Francis Steegmuller. Cambridge: Harvard University Press, 1982.

———. *Carnets de travail*. Critical and genetic edition, edited by Pierre-Marc de Biasi. Paris: Balland, 1988.

———. *A Simple Heart*. Translated by Charlotte Mandell. Hoboken, N.J.: Melville House, 2004.

Floren, Josef. *Studien zur Typologie des Gorgoneion*. Munich: Aschendorff, 1977.

Foucault, Michel. *The Archeology of Knowledge*. Translated by A. M. Sheridan Smith. New York: Pantheon, 1972.

———. *Discipline and Punish: The Birth of the Prison*. Translated by Alan Sheridan. New York: Vintage, 1979.

———. "Why Study Power: The Question of the Subject." In *Beyond Structuralism and Hermeneutics*, edited by Hubert L. Dreyfus and Paul Rabinow, 208–16. Chicago: University of Chicago Press, 1982.

———. "Society Must Be Defended." In *Ethics: Subjectivity and Truth*, edited by Paul Rabinow, 59–65. New York: New Press, 1997.

———. "Of Other Spaces." In *The Visual Culture Reader*, edited by Nicholas Mirzoeff, 237–44. London and New York: Routledge, 1998.

Francisco Pacheco, Joaquín, Francisco de Cárdenas y Espejo, and Luis Torres de Mendoza, eds. *Colección de documentos ineditos relativos al descubrimiento, conquista y organización de las posesiones españolas en América y Oceanía*. 24 vols. First series. Madrid: Imprenta de M. Bernaldo de Quirós, 1864–84.

Fraser, David. "Joseph Wright of Derby et la Lunar Society: Essai sur les rapports de l'artiste avec la science et l'industrie." In *Joseph Wright of Derby, 1734–1797*, edited by Judy Egerton. Exhibition catalog. Paris: Réunion des Musées Nationaux, 1990.

Freud, Sigmund. "From the History of an Infantile Neurosis." In *An Infantile Neurosis and Other Works*. Vol. 17 of *The Standard Edition of the Complete Psychological Works of Sigmund Freud*, edited by James Strachey. London: Hogarth, 1964.

Furttenbach, Joseph. *Architectura navalis*. Ulm, 1629; reprinted, Hildesheim and New York: Olms, 1975.

Geib, George W. "The Land Ordinance of 1785: A Bicentennial Review." *Indiana Magazine of History* 81, no. 1 (1985): 1–13.

Gengenbach, Pamphilus [pseud.]. *Von der falschen Betler buberey [Liber vagatorum]*. With an introduction by Martin Luther. Wittemberg and Magdeburg: Öttinger, 1528.

Gennep, Arnold van. *The Rites of Passage*. Translated by Monika Vizedom and Gabrielle L. Caffee. London: Routledge, 1977.

Geremek, Bronislaw. *Poverty: A History*. Translated by Agnieszka Kolakowska. Oxford: Blackwell, 1994.

Gifford, Philip C. "Trait Origins in Trobriand War-Shields: The Uncommon Selection of an Image Cluster." *Anthropological Papers of the American Museum of Natural History* 79 (1996): 2–13.

Glasmeier, Michael. "Ansichten von Zeichnungen." In *Entwerfen und Entwurf: Praxis und Theorie des künstlerischen Schaffensprozesses*. Edited by Gundel Mattenklott and Friedrich Weltzien, 75–86. Berlin: Reimer, 2003.

Gogwilt, Christopher. *The Invention of the West: Joseph Conrad and the Double-Mapping of Europe and the Empire*. Stanford: Stanford University Press, 1995.

Gonzáles García, Pedro, et al. *Archivo General de Indias: Archivos españole*. Barcelona: Lunwerg Editores; Madrid: Ministerio de Cultura, Dirección General del Libro, Archivos y Bibliotecas, 1995.

Granzow, Uwe. *Quadrant, Kompass und Chronometer: Technische Implikationen des euro–asiatischen Seehandels von 1500 bis 1800*. Stuttgart: Steiner, 1986.

Graves, Robert. *The Greek Myths*. 2 vols. Rev. ed. Harmondsworth: Penguin, 1986.

Griaule, Marcel. "Seuil." In *Documents: Archéologie, Beaux-Arts, Ethnographie, Variétés* 2, no. 2 (1930): 103.

Grimm, Claus, ed. *Stilleben: Die italienischen, spanischen und französischen Meister*. Stuttgart: Belser, 1995.

Groebner, Valentin. *Who Are You? Identification, Deception, and Surveillance in Early Modern Europe*. Translated by Mark Kyburz and John Peck. New York: Zone Books, 2007.

Grotius, Hugo. *The Free Sea*. Translated by Richard Hakluyt and edited by David Armitage. Indianapolis: Liberty Fund, 2004.

Guidoni, Enrico, and Angela Marino Guidoni. *Storia dell'urbanistica: Il cinquecento*. Bari: Laterza, 1982.

Gumbrecht, Hans Ulrich. "Flache Diskurse." In *Materialität der Kommunikation*, edited by Hans Ulrich Gumbrecht and K. Ludwig Pfeiffer, 914–23. Frankfurt/M: Suhrkamp, 1988.

———. "A Farewell to Interpretation." In *Materialities of Communication*, edited by Hans Ulrich Gumbrecht and K. Ludwig Pfeiffer and translated by William Whobrey, 389–402. Stanford: Stanford University Press, 1994.

Gutton, Jean-Pierre. *La société et les pauvres en Europe, XVIe-XVIIIe siècles.* Paris: Presses Universitaires de France, 1974.

Hamacher, Werner. *Pleroma: Reading in Hegel; The Genesis and Structure of a Dialectical Hermeneutics in Hegel.* Translated by Nicholas Walker and Simon Jarvis. London: Athlone, 1998.

Hardoy, Jorge Enrique. *Cartografía urbana colonial de América Latina y el Caribe.* Buenos Aires: Grupo Editor Latinoamericano, 1991.

Harley, John B., and David Woodward, eds. *Cartography in Prehistoric, Ancient, and Medieval Europe and the Mediterranean.* Vol. 1 of *The History of Cartography.* Chicago: University of Chicago Press, 1987.

Harrus-Révidi, Gisèle. *Die Kunst des Geniessens: Esskultur und Lebenslust.* Translated by Renate Sandner and Thorsten Schmidt. Düsseldorf and Zürich: Artemis und Winkler, 1996.

Hauschild, Thomas. *Der böse Blick: Ideengeschichtliche und sozialpsychologische Untersuchungen.* 2nd ed. Berlin: Verlag Mensch und Leben, 1982.

Hayles, N. Kathrine. *How We Became Posthuman: Virtual Bodies in Cybernetics, Literature, and Informatics.* Chicago: University of Chicago Press, 1999.

———. "Cybernetics." In *Critical Terms for Media Studies,* edited by William J. T. Mitchell and Mark B. N. Hansen, 145–56. Chicago: University of Chicago Press, 2010.

Hedio, Caspar. "Der Bericht Hedios." In *Das Marburger Religionsgespräch 1529,* edited by Gerhard May, 13–32. Gütersloh, Germany: Mohn, 1970.

Hegel, Gottfried Wilhelm Friedrich. "The Spirit of Christianity." In *Early Theological Writings,* translated by T. M. Knox. Philadelphia: University of Pennsylvania Press, 1971.

Heidegger, Martin. *Introduction to Metaphysics.* Translated by Ralph Manheim. New Haven: Yale University Press, 1959.

———. *Being and Time.* Translated by John Macquarie and Edward Robinson. London: SCM Press, 1962.

———. "The Age of the World Picture." In *The Question Concerning Technology and Other Essays,* translated and with an introduction by William Lovitt, 115–54. New York: Harper and Row, 1977.

———. "Time and Being." In *On Time and Being,* translated by Joan Stambaugh, 1–25. Chicago: University of Chicago Press, 2002.

Hendrix, Lee, and Thea Vignau-Wilberg, eds. *Mira calligraphiae monumenta: A Sixteenth-Century Calligraphic Manuscript Inscribed by Georg Bocskay and Illuminated by Joris Hoefnagel.* Los Angeles: J. Paul Getty Museum, 1992.

Herder, Johann Gottfried von. *Sämtliche Werke.* Edited by Bernhard Suphan. 33 vols. Berlin: Weidmann, 1877–1913.

———. "Vitae non scholae discendum." *Sämtliche Werke* 30:266–74.

———. "Von der Ausbildung der Rede und Sprache in Kindern und Jünglingen." *Sämtliche Werke* 30:217–26.

———. "Treatise on the Origin of Language (1772)." In *Herder: Philosophical Writings,* translated and edited by Michael N. Forster, 65–166. Cambridge: Cambridge University Press, 2002.

Hesiod. *"Theogony" and "Works and Days."* Translated by M. L. West. Oxford: Oxford University Press, 1989.

Hiltbrunner, Otto. *Gastfreundschaft in der Antike und im frühen Christentum.* Darmstadt: Wissenschaftliche Buchgesellschaft, 2005.

Hiltbrunner, Otto, Denys Gorce, and Hans Wehr. "Gastfreundschaft." In Klauser et al., eds., *Reallexikon für Antike und Christentum,* vol. 8, columns 1061–123.

Hörisch, Jochen. *Brot und Wein: Die Poesie des Abendmahls.* Frankfurt/M: Suhrkamp, 1992.

Hoffmann, Volker. "Filippo Brunelleschi: Kuppelbau und Perspektive." In *Saggi in onore di Renato Bonelli: Quaderni dell'istituto di storia dell'architettura,* edited by Corrado Bozzoni, Giovanni Carbonara, and Gabriela Villetti, 317–26. Rome: Multigrafica Editrice, 1992.

Holl, Adolf. "Das erste Letzte Abendmahl." In *Speisen, Schlemmen, Fasten: Eine Kulturgeschichte des Essens,* edited by Uwe Schultz, 43–55. Frankfurt/M: Insel, 1993.

Holl, Ute. "Postkoloniale Resonanzen." *Archiv für Mediengeschichte* 11 (2011): 115–28.

Horn, Eva. "'There Are No Media.'" *Grey Room* 29 (2007): 6–13.

Houston, James M. "The Foundation of Colonial Towns in Hispanic America." In *Urbanization and Its Problems: Essays in Honour of E. W. Gilbert,* edited by Robert P. Beckinsale and James M. Houston, 352–90. Oxford: Blackwell, 1968.

Ingold, Tim. *Lines: A Brief History.* London and New York: Routledge: 2007.

———. "Toward an Ecology of Materials." *Annual Review of Anthropology* 41 (2012): 427–42.

Jakobson, Roman. "Linguistics and Poetics." In *Style in Language,* edited by Thomas Sebeok, 350–77. Cambridge: MIT Press, 1960.

———. "Shifters, Verbal Categories, and the Russian Verb." In *Selected Writings,* vol. 2, *World and Language,* 130–47. The Hague: Mouton, 1971.

Janson, Horst Waldemar. *Apes and Ape Lore in the Middle Ages and the Renaissance.* London: Warburg Institute, 1952.

Johnson, Hildegard Binder. *Order upon the Land: The U.S. Rectangular Land Survey and the Upper Mississippi Country.* New York: Oxford University Press, 1976.

Johnston, Stephen. "Making Mathematical Practice: Gentlemen, Practitioners and Artisans in Elizabethan England." Ph.D. diss., Cambridge University, 1994.

Kafka, Franz. *The Castle.* Translated by Willa Muir and Edwin Muir. London: Secker and Warburg, 1957.

———. *Letters to Felice.* Edited by Erich Heller and Jürgen Born and translated by James Stern and Elisabeth Duckworth. New York: Schocken, 1973.

———. "Before the Law." In *"The Metamorphosis," "In the Penal Colony," and Other Stories,* translated by Joachim Neugroschel, 148–49. New York: Scribner, 1995.

Kant, Immanuel. "Eternal Peace." In *The Philosophy of Kant: Immanuel Kant's Moral and Political Writings,* edited by Carl J. Friedrich. New York: Modern Library, 1949.

———. *Religion within the Bounds of Bare Reason.* Translated by Werner Pluhar. Indianapolis: Hackett, 2009.

Kantorowicz, Ernst H. *The King's Two Bodies: A Study in Mediaeval Political Theology.* Princeton: Princeton University Press, 1957.

Kathan, Bernhard. *Zum Fressen gern: Zwischen Haustier und Schlachtvieh*. Berlin: Kadmos, 2004.

Kaufmann, Stefan. *Soziologie der Landschaft*. Wiesbaden: VS Verlag für Sozialwissenschaften, 2005.

Kaufmann, Thomas DaCosta, and Virginia Roehrig Kaufmann. "The Sanctification of Nature: Observations on the Origins of Trompe l'Oeil in Netherlandish Book Painting of the Fifteenth and Sixteenth Centuries." *J. Paul Getty Museum Journal* 19 (1991): 43–64.

Kemp, Martin. *The Science of Art: Optical Themes in Western Art from Brunelleschi to Seurat*. New Haven: Yale University Press, 1990.

Kemp, Wolfgang. "Disegno: Beiträge zur Geschichte des Begriffs zwischen 1547 und 1607." *Marburger Jahrbuch* 19 (1974): 219–40.

Kish, George. "Early Thematic Mapping: The Work of Phillippe Buache." *Imago Mundi* 28 (1976): 129–36.

Kittler, Friedrich A. "Autorschaft und Liebe." In *Austreibung des Geistes aus den Geisteswissenschaften: Programme des Poststrukturalismus*, edited by Friedrich A. Kittler, 142–73. Paderborn: Schöningh, 1980.

———. *Discourse Networks, 1800/1900*. Translated by Michael Metteer and Chris Cullens. Stanford: Stanford University Press, 1990.

———. "The World of the Symbolic—A World of the Machine." In *Literature, Media, Information Systems: Essays*, edited by John Johnston and translated by Stephanie Harris, 130–46. Amsterdam: OAP, 1997.

———. *Unsterbliche*. Munich: Fink, 2004.

Klausman, Theodor, et al., eds. *Reallexikon für Antike und Christentum: Sachwörterbuch zur Auseinandersetzung des Christentums mit der antiken Welt*. 22 vols. Stuttgart: Hiersemann, 1950–.

Klein, Wolf Peter, and Marthe Grund. "Die Geschichte der Auslassungspunkte: Zu Entstehung, Form und Funktion der deutschen Interpunktion." *Zeitschrift für germanistische Linguistik* 25, no. 1 (1997), 24–44.

Kleist, Heinrich von. *The Feud of the Schroffensteins*. Translated by Mary J. Price and Lawrence M. Price. *Poet Lore* 27, no. 5 (Sept.-Oct. 1916).

———. *Penthesilea*. In *Five Plays*, translated by Martin Greenberg. New Haven: Yale University Press, 1988.

Klose, Alexander. "Vom Raster zur Zelle—die Containerisierung der modernen Architektur." Lecture delivered at Bauhaus University, Weimar, January 17, 2006. Delivered in English as "From Grid to Box: The Containerization of Modern Architecture" at the Goethe Institute, Prague, October 7, 2005, at the workshop "City-Media-Space." Both versions available at http://www.containerwelt.info./ordner_eigene_texte.html.

———. *Das Container-Prinzip: Wie eine Box unser Denken verändert*. Hamburg: Mare, 2009.

Klusemann, Kurt. *Das Bauopfer: Eine ethnographisch-prähistorisch-linguistische Studie*. Graz-Hamburg: Selbstverlag, 1919.

Knies, Karl. *Der Telegraph als Verkehrsmittel: Mit Erörterungen über den Nachrichtenverkehr überhaupt*. Tübingen: Laupp, 1857.

König, Eberhard, and Christiane Schön, eds. *Stilleben*. Vol. 5 of *Geschichte der klassische Bildgattungen in Quellentexten und Kommentaren*. Berlin: Dietrich Reimer Verlag, 1996.

Konvitz, Josef W. *Cartography in France, 1660–1848: Science, Engineering, and Statecraft*. Chicago: University of Chicago Press, 1987.

Kraatz, August. *Maschinentelegraphen*. Telegraphen- und Fernsprechtechnik in Einzeldarstellungen, ed. Theodor Karras, no. 1. Braunschweig: Friedrich Vieweg, 1906.

Krauss, Rosalind. "Grids." *October* 9 (1979): 50–64.

Krausse, Joachim. "Information at a Glance: On the History of the Diagram." *OASE: Tijdschrift voor architectuur* 48 (1998): 3–30.

Krausse, Joachim, and Claude Lichtenstein. "Earthwalking—Skyriding: An Invitation to Join R. Buckminster Fuller on a Voyage of Discovery." In *Your Private Sky: R. Buckminster Fuller: Discourse*, by R. Buckminster Fuller et al., edited by Joachim Krausse and Claude Lichtenstein, 7–43. Baden and Zürich: Springer, 2001.

Kren, Thomas. "Revolution and Transformation: Painting in Devotional Manuscripts, circa 1467–1485." In *Illuminating the Renaissance: The Triumph of Flemish Manuscript Painting in Europe*, edited by Thomas Kren and Scot McKendrick, 121–221. Los Angeles: J. Paul Getty Museum, 2003.

Kretzenbacher, Leopold. *Santa Lucia und die Lutzelfrau: Volksglaube und Hochreligion im Spannungsfeld Mittel- und Südeuropas*. Munich: Oldenbourg, 1959.

Lacan, Jacques. "7ème Congrès de l'École freudienne de Paris à Rome." *Les Lettres de l'École freudienne*, no. 16 (1975): 177–203. Published in English as "Press Conference by Doctor Jacques Lacan at the French Cultural Center, Rome, 29-October-1974," trans. Richard G. Klein, http://www.freud2lacan.com/docs/lacan_press.pdf (accessed 10 July 2012).

———. *The Seminar of Jacques Lacan, Book 2: The Ego in Freud's Theory and in the Technique of Psychoanalysis, 1954–1955*. Translated by Sylvana Tomaselli, with notes by John Forrester. Cambridge: Cambridge University Press, 1988.

———. *The Seminar of Jacques Lacan, Book 3: The Psychoses, 1955–1956*. Edited by Jacques-Alain Miller and translated by Russell Grigg. New York: Norton, 1997.

———. *The Seminar of Jacques Lacan, Book 11: The Four Fundamental Concepts of Psychoanalysis*. Edited by Jacques-Alain Miller and translated by Alan Sheridan. New York: Norton, 1998.

———. *Le séminaire de Jacques Lacan, livre 10: L'angoisse*. Edited by Jacques-Alain Miller. Paris: Seuil, 2004.

———. *Le triomphe de la religion*. Paris: Editions du Seuil, 2005.

Latour, Bruno. "Drawing Things Together." In *Representation in Scientific Practice*, edited by Michael Lynch and Steve Woolgar, 19–68. Cambridge: MIT Press, 1990.

———. "The 'Pédofil' of Boa Vista: A Photo-Philosophical Montage." *Common Knowledge* 4, no. 1 (1995): 144–87.

Latour, Bruno, and Steve Woolgar. *Laboratory Life: The Construction of Scientific Facts*. Princeton: Princeton University Press, 1986.

Le Corbusier. "American Prologue." In *Precisions on the Present State of Architecture and City Planning*, translated by Edith Schreiber Aujame. Cambridge: MIT Press, 1991.

———. "A Dwelling at Human Scale." In *Precisions on the Present State of Architecture and City Planning*, translated by Edith Schreiber Aujame, 85–103. Cambridge: MIT Press, 1991.

Leach, Edmund R. "A Trobriand Medusa?" *Man* 54 (July 1954): 103–05.

———. "'A Trobriand Medusa?' A Reply to Dr. Berndt." *Man* 58 (May 1958), 79.

Lebrun, Pierre. "Recueil des essaies des merveilles de la peinture (1635)." In *Medieval and Renaissance Treatises on the Arts of Painting: Original Texts with English Translations*, edited by Mary Philadelphia Merrifield, 767–841. London: John Murray, 1849; reprinted, Mineola, N.Y.: Dover Publications, 1999.

Leclercq, Jean. "Saint Bernard et ses sécrétaires." *Révue bénédictine* 61 (1951): 208–29.

Lee, Bertram, ed. *Libros de Cabildos de Lima*, vol. 2. Lima: Torres Aguirre, 1935.

Leibniz, Gottfried Wilhelm. *Fragmente zur Logik*. Edited by Franz Schmidt. Berlin: Akademie Verlag, 1960.

Leonhard, Karin, and Robert Felfe. *Lochmuster und Linienspiel: Überlegungen zur Druckgrafik des 17. Jahrhunderts*. Freiburg im Breisgau and Berlin: Rombach, 2006.

Leonardo da Vinci. *The Notebooks of Leonardo da Vinci*. Edited and translated by Edward MacCurdy. 2 vols. New York: Reynal and Hitchcock, 1938.

———. *Das Wasserbuch: Schriften und Zeichnungen*. Translated by Marianne Schneider. Munich: Schirmer/Mosel, 1996.

Leroi-Gourhan, André. *Treasures of Prehistoric Art*. Translated by Norbert Guterman. New York: Abrams, 1967.

Lévi-Strauss, Claude. *The Savage Mind*. London: Weidenfeld and Nicolson, 1972.

———. *The Origin of Table Manners*. Vol. 3 of *Introduction to a Science of Mythology*. Translated by John Weightmann and Doreen Weightmann. New York: Harper and Row, 1978.

Loo, Hulin de. "La vignette chez les enlumineurs gantois entre 1470 et 1500." *Bulletin de la classe de Beaux Arts de L'Academie Royale de Belgique* 21 (1939): 158–80.

López de Pinero, José Maria. *El arte de navegar en la España del Rinacimiento*. Barcelona: Editorial Labor, 1986.

Lovink, Geert. "Whereabouts of German Media Theory." In *Zero Comment: Blogging and Critical Internet Culture*, 83–98. New York and London: Routledge, 2008.

Luchs, Alison. *The Mermaids of Venice: Fantastic Sea Creatures in Venetian Renaissance Art*. London: Harvey Miller, 2010.

Lutz, Helga. "Medien des Entbergens: Falt- und Klappoperationen in der altniederländischen Kunst des späten 14. und frühen 15. Jahrhunderts." *Archiv für Mediengeschichte* 10 (2010): 27–46.

Lyotard, Jean-François. *Discourse, Figure*. Translated by Antony Hudek and Mary Lydon. Minneapolis: University of Minnesota Press, 2011.

McGee, David. "From Craftsmanship to Draftsmanship: Naval Architecture and the Three Traditions of Early Modern Design." *Technology and Culture* 40, no. 2 (1999): 209–36.

Maccarrone, Michele. *Vicarius Christi: Storia del titolo papale*. Lateranum, new series 18. Rome: Facultas Theologica Pontificii Athenaei Lateranensis, 1952.

Macho, Thomas. "Tier." In *Vom Menschen: Handbuch Historische Anthropologie*, edited by Christoph Wulf, 62–85. Weinheim-Basel: Beltz, 1997.

———. "Glossolalie in der Theologie." In *Zwischen Rauschen und Offenbarung: Zur Kultur- und Mediengeschichte der Stimme*, edited by Friedrich Kittler, Thomas Macho, and Sigrid Weigel, 3–17. Berlin: Akademie Verlag, 2002.

———. "Tieropfer: Zur Geschichte der rituellen Tötung von Tieren." In *Mensch und Tier: Eine paradoxe Beziehung*, edited by Stiftung Deutsches Hygiene-Museum, 54–69. Ostfildern-Ruit, Germany: Hatje Cantz, 2002.

———. "Zeit und Zahl: Kalender- und Zeitrechnung als Kulturtechniken." In *Bild—Schrift—Zahl*, edited by Sybille Krämer and Horst Bredekamp, 179–92. Munich: Wilhelm Fink Verlag, 2003.

———. "Tiere zweiter Ordnung: Kulturtechniken der Identität und Identifikation." In *Über Kultur: Theorie und Praxis der Kulturreflexion*, edited by Dirk Baecker et al., 99–117. Bielefeld: Transcript, 2008.

———. *Vorbilder*. Munich: Wilhelm Fink Verlag, 2011.

———. "Second-Order Animals: Cultural Techniques of Identity and Identification." *Theory, Culture and Society* 30, no. 6 (2013): 31.

Malinowski, Bronisław. "The Problem of Meaning in Primitive Languages." Supplement to *The Meaning of Meaning: A Study of the Influence of Language upon Thought and of the Science of Symbolism*, by C. K. Ogden and I. A. Richards, 296-336. London: Routledge and Kegan Paul, 1949.

———. *Argonauts of the Western Pacific: An Account of Native Enterprise and Adventure in the Archipelagos of Melanesian New Guinea*. With an introduction by James G. Fraser. New York: E. P. Dutton, 1961.

Marin, Louis. "Representation and Simulacrum." In *On Representation*, translated by Catherine Porter, 309–19. Stanford: Stanford University Press, 2001.

Marót, Károly. *Die Anfänge der griechischen Literatur: Vorfragen*. Budapest: Verlag der Ungarischen Akademie der Wissenschaften, 1960.

Marrow, James H. "Scholarship on Flemish Manuscript Illumination of the Renaissance: Remarks on Past, Present, and Future." In *Flemish Manuscript Painting in Context: Recent Research*, edited by Elizabeth Morrison and Thomas Kren, 163–76. Los Angeles: Getty Publications, 2006.

Martin, Norman F. *Los vagabundos en la Nueva Espana, siglo XVI*. Mexico City: Editorial Jus, 1957.

Mattenklott, Gundel, and Friedrich Weltzien, eds. *Entwerfen und Entwurf: Praxis und Theorie des künstlerischen Schaffensprozesses*. Berlin: Reimer, 2003.

Maumi, Catherine. *Thomas Jefferson et le projet du Nouveau Monde*. Paris: Éditions de la Villette, 2007.

Mauss, Marcel. "Techniques of the Body." In *Incorporations*, edited by Jonathan Carey and Sanford Kwinter, 454–77. New York: Zone Books, 1992.

Maxwell, James Clerk. *The Scientific Letters and Papers of James Clerk Maxwell*. Edited by P. M. Harman. Cambridge: Cambridge University Press, 1995.

Maye, Harun. "Was ist eine Kulturtechnik?" *Zeitschrift für Medien- und Kulturforschung* no. 1 (2010): 121–36.

Mies van der Rohe, Ludwig. "Hochhausprojekt für Bahnhof Friedrichstrasse in Berlin (1922)." In *Frühlicht 1920–1922: Eine Folge für die Verwirklichung des neuen Baugedankens,* edited by Bruno Taut, 212–14. Bauwelt Fundamente 8. Berlin, Frankfurt/M., and Vienna: Ullstein, 1963

Mommsen, Theodor, ed. *Res gestae Divi Augusti: Ex monumentis Ancyrano et Apolloniensi iterum.* Berlin: apud Weidmannos, 1865.

Monmonier, Mark. *Rhumb Lines and Map Wars: A Social History of the Mercator Projection.* Chicago: University of Chicago Press, 2004.

Morrison, John S., and Roderick Trevor Williams. *Greek Oared Ships, 900–322 B.C.* Cambridge: Cambridge University Press, 1968.

Musil, Robert. "Doors and Portals." In *Posthumous Papers of a Living Author,* translated by Peter Wortsman, 59–63. Hygiene, Colo.: Eridanos, 1987.

Nash, Susie. *Northern Renaissance Art.* Oxford and New York: Oxford University Press, 2008.

Neufert, Ernst. *Bauordnungslehre: Handbuch für rationelles Bauen nach geregeltem Mass.* 1943; reprinted, Frankfurt/M. and Berlin: Ullstein, 1961. Published in English as *Architect's Data*; first English edition edited and revised by Rudolf Herz from translations by G. H. Berger et al. (London: Lockwood, 1970).

Neumann, Gerhard. "Nachrichten vom 'Pontus': Das Problem der Kunst im Werk Franz Kafkas." In *Franz Kafka Symposion 1983,* edited by Wilhelm Emrich and Bernd Goldmann, 101–57. Mainz: Hase und Koehler, 1985.

———. "Hexenküche und Abendmahl: Die Sprache der Liebe im Werk Heinrich von Kleists." *Freiburger Universitätsblätter* 91 (1986): 9–31.

———. " 'Jede Nahrung ist ein Symbol': Umrisse einer Kulturwissenschaft des Essens." In *Kulturthema Essen: Ansichten und Problemfelder,* edited by Alois Wierlacher, Gerhard Neumann, and Hans J. Teuteberg, 385–444. Berlin: Akademie Verlag, 1993.

———. "Tania Blixen: 'Babettes Gastmahl.' " In *Kulturthema Essen: Ansichten und Problemfelder,* edited by Alois Wierlacher, Gerhard Neumann, and Hans J. Teuteberg, 289–318. Berlin: Akademie Verlag, 1993.

———. "Der Blick des Anderen: Zum Motiv des Hundes und des Affen in der Literatur." *Jahrbuch der deutschen Schillergesellschaft* 40 (1996): 87–122.

Nowak, Troy Joseph. "Archaeological Evidence for Ship Eyes: An Analysis of Their Form and Function." Master's thesis, Texas A&M University, 2006.

Nueva Colección de documentos para la historia de México. Mexico City: Editorial Salvador Chávez Hayhoe, 1886–92.

Oberhuber, Florian. "Der Vagabund." In *Grenzverletzer: Von Schmugglern, Spionen und anderen subversiven Gestalten,* edited by Eva Horn, Stefan Kaufmann, and Ulrich Bröckling, 58–79. Berlin: Kadmos, 2002.

Ong, Walter J. *Orality and Literacy: The Technologizing of the Word.* London and New York: Methuen, 1982.

Orth, Myra D. "What Goes Around: Borders and Frames in French Manuscripts." In "Essays in Honor of Lilian M. C. Randall," special issue, *Journal of the Walters Art Gallery* 54 (1996): 189–203.

Osiander, Andreas. "Bericht Osianders an den Nürnberger Rat." In *Das Marburger Religionsgespräch 1529*, edited by Gerhard May. Gütersloh, Germany: Mohn, 1970.

Ottomeyer, Hans. "Tischgerät und Tafelbräuche: Die Kunstgeschichte als Beitrag zur Kulturforschung des Essens." In *Kulturthema Essen: Ansichten und Problemfelder*, edited by Alois Wierlacher, Gerhard Neumann, and Hans J. Teuteberg, 177–85. Berlin: Akademie Verlag, 1993.

Paczensky, Gert von, and Anna. Dünnebier. *Leere Töpfe, volle Töpfe: Die Kulturgeschichte des Essens und Trinkens*. Munich: Knaus, 1994.

Pächt, Otto. *The Master of Mary of Burgundy*. London: Faber and Faber, 1948.

Pattison, William D. *Beginnings of the American Rectangular Land Survey System, 1784–1800*. New York: Arno Press, 1979.

Pestalozzi, Johann Heinrich. "Über den Sinn des Gehörs in Hinsicht auf Menschenbildung durch Ton und Sprache." In *Ausgewählte Schriften*, edited by Wilhelm Flitner, 246–270. Frankfurt/M.: Ullstein, 1983.

Peters, John. "Strange Sympathies: Horizons of German and American Media Theory." In *American Studies as Media Studies*, edited by Frank Kelleter and Daniel Stein, 3–23. Heidelberg: Winter, 2008.

Pett, Phineas. *The Autobiography of Phineas Pett*. Edited by W. G. Perrin. Publications of the Navy Records Society, vol. 51. London: Ballantyne Press, 1918.

Pietzcker, Achim. "Schiebetür." *Arch+: Zeitschrift für Architektur und Städtebau* 191/192 (March 2009): 91.

Plato. *Timaeus*. Translated by Donald J. Zeyl. Indianapolis: Hackett Publishing, 2000.

Pliny the Elder, Gaius. *The Natural History of Pliny*, vol. 2. Translated by John Bostock and H. T. Riley. London: George Bell, 1890.

Prigge, Walter. "Typologie und Norm: Zum modernen Traum der industriellen Fertigung von Wohnungen." In *Constructing Utopia: Konstruktionen künstlicher Welten*, edited by Annett Zinsmeister, 69–77. Berlin and Zürich: Diaphanes, 2005.

Pulido Rubio, José. *El Piloto mayor de la Casa de la Contratación de Sevilla: Pilotos mayores, Catedraticos de Cosmografia y Cosmografos*. Seville: Escuela de Estudios Hispano-Americanos, 1950.

Pulitzer Los, Natasha. "The Culture of Ship Design." *Rassegna: Themes in Architecture* 12, no. 44/4 (1990): 9–12.

Pynchon, Thomas. "Mortality and Mercy in Vienna." *Epoch* 9, no. 4 (1959): 195–213.

———. "Entropy." In *Slow Learner*, 79–98. Boston and Toronto: Little, Brown, 1984.

Queller, Donald E. *The Office of Ambassador in the Middle Ages*. Princeton: Princeton University Press, 1967.

Rama, Angel. *The Lettered City*. Translated by John Charles Chasteen. Durham, N.C.: Duke University Press, 1996.

Randall, Lilian M. C. *Images in the Margins of Gothic Manuscripts*. Berkeley and Los Angeles: University of California Press, 1966.

Rath, Claus Dieter. *Reste der Tafelrunde: Das Abenteuer der Esskultur.* Reinbek: Rowohlt, 1984.

Real Academia de la Historia, ed. *Las siete Partidas del Rey Don Alfonso el Sabio, cotejadas con varios codices antiguos.* Vol. 1. Madrid: Imprenta Real, 1807.

Real Ramos, César. "'Fingierte Armut' als Obsession und die Geburt des auktorialen Erzählers in der Picaresca." In *Der Ursprung von Literatur: Medien, Rollen, Kommunikationssituationen zwischen 1450 und 1650,* edited by Gisela Smolka-Koerdt, Peter M. Spangenberg and Dagmar Tillmann-Bartylla, 175–90. Munich: Wilhelm Fink Verlag, 1988.

Recopilación de leyes de los reynos de las Indias. 5 vols. Madrid, 1681; reprinted, Mexico City: M. A. Porrúa, 1987.

Rheinberger, Hans-Jörg. *Toward a History of Epistemic Things: Synthesizing Proteins in the Test Tube.* Stanford: Stanford University Press, 1997.

Rhodius, Apollonius. *Argonautica.* Edited and translated by William H. Race. Loeb Classical Library. Cambridge: Harvard University Press, 2008.

Riepe, Manfred. *Bildgeschwüre: Körper und Fremdkörper im Kino David Cronenbergs: Psychoanalytische Filmlektüren nach Freud und Lacan.* Bielefeld: transcript, 2002.

Rieth, Eric. "Remarques sur une série d'illustrations de l'*Ars Nautica* (1570) de Fernando Oliveira." *Neptunia* 169 (1988): 36–43.

Rotman, Brian. *Signifying Nothing: The Semiotics of Zero.* Stanford: Stanford California Press, 1993.

Rousseau, Jean-Jacques. *Politics and the Arts: Letter to M. d'Alembert on the Theatre.* Translated by Allen Bloom. Glencoe, Ill.: Free Press, 1960.

———. *Confessions.* Translated by Angela Scholar. Oxford: Oxford University Press, 2000.

Rubin, Miri. *Gentile Tales: The Narrative Assault on Late Medieval Jews.* New Haven: Yale University Press, 1999.

Russell, Margarita. *Visions of the Sea: Hendrick C. Vroom and the Origins of Dutch Marine Painting.* Leiden: Brill and Leiden University Press, 1983.

Salisbury, Richard F. "Trobriand Medusa?" *Man* 59 (March 1959): 50–51.

Sander, Jochen, ed. *Die Magie der Dinge: Stillebenmalerei 1500–1800.* Exhibition catalog of the Städel Museum, Frankfurt a. M., and Kunstmuseum Basel. Ostfildern, Germany: Hatje Cantz, 2008.

Schade, Karl. *Ad excitandum devotionis affectum: Kleine Triptychen in der altniederländischen Malerei.* Weimar: VDG, 2001.

Schäffner, Wolfgang. "Punto, línea, abertura: Elementos para una historia medial de la arquitectura y el diseño." Walter Gropius Chair Inaugural Lecture Series, 2003/2004, FADU, University of Buenos Aires. Typescript in possession of Bernhard Siegert.

Schmidgen, Henning. *Hirn und Zeit: Die Geschichte eines Experiments 1800–1950.* Berlin: Matthes und Seitz, forthcoming.

Schmitt, Carl. *The Nomos of the Earth in the International Law of the Jus Publicum Europaeum.* Translated by G. L. Ulmen. New York: Telos Press, 2003.

Schneider, Birgit. *Textiles Prozessieren: Eine Mediengeschichte der Lochkartenweberei.* Zürich and Berlin: Diaphanes, 2007.

Schneider, Manfred. *Liebe und Betrug: Die Sprachen des Verlangens.* Munich and Vienna: Carl Hanser Verlag, 1992.

———. "Luther with McLuhan." In *Religion and Media*, edited by Hent de Vries and Samuel Weber, 198–215. Stanford: Stanford University Press, 2001.

———. "Das Notariat der Hunde: Eine literaturwissenschaftliche Kynologie." *Zeitschrift für deutsche Philologie* 126 (2007): 4–27.

Schneider, Ute. *Die Macht der Karten: Eine Geschichte der Kartographie vom Mittelalter bis heute.* Darmstadt: Primus Verlag, 2004.

Schott, Andreas. "Ampliss: Viro Augerio Busbequio Exlegato Byzantino, & supremo Curiae Isabellae Praefecto" [Dedication]. In Sextus Aurelius Victor, *De vita et moribus imperatorum romanorum: Excerpta ex libris Sexti Aurelii Victoris, à Caesare Augusto usque ad Theodosium imperatorem*, edited by Andreas Schott, 3–7. Antwerp: Ex officina Christophori Plantini, 1579.

Schüttpelz, Erhard. "Die medienanthropologische Kehre der Kulturtechniken." *Archiv für Mediengeschichte* 6 (2006): 87–110.

Schupbach, William. "A Select Iconography of Animal Experiment." In *Vivisection in Historical Perspective*, edited by Nicolaas Rupke. London: Croom Helm, 1987.

Semper, Gottfried. *Style in the Technical and Tectonic Arts; or, Practical Aesthetics.* Translated by Harry Francis Mallgrave and Michael Robinson. Los Angeles: Getty Research Institute, 2004.

Serres, Michel. "Turner traduit Carnot." In *Hermes III: La Traduction,* 233–42. Paris: Éditions de Minuit, 1974.

———. *The Parasite.* Translated by Lawrence R. Schehr. Baltimore: Johns Hopkins University Press, 1982. Originally published as *Le parasite* (Paris: B. Grasset, 1980).

———."The Origin of Language." In *Hermes: Literature, Science, Philosophy*, edited by Josué V. Harari and David F. Bell. Baltimore: Johns Hopkins University Press, 1992.

———. "Platonic Dialogue." In *Hermes: Literature, Science, Philosophy*, edited by Josué V. Harari and David F. Bell. Baltimore: Johns Hopkins University Press, 1992.

Shannon, Claude Elwood. "A Mathematical Theory of Communication." In *Claude Elwood Shannon: Collected Papers*, edited by N. J. A. Sloan and Aaron D. Wyner. Piscataway, N.J.: IEEE Press, 1993.

Siegert, Bernhard. *Relays: Literature as an Epoch of the Postal System.* Translated by Kevin Repp. Stanford: Stanford University Press, 1999.

———. *Passage des Digitalen. Zeichenpraktiken der neuzeitlichen Wissenshaften 1500–1900.* Berlin: Brinkmann und Bose, 2003.

———.*Passagiere und Papiere: Schreibakte auf der Schwelle zwischen Spanien und Amerika.* Munich: Wilhelm Fink Verlag, 2006.

———. "After the Wall: Interferences among Grids and Veils." *Graz Architektur Magazin* 9 (2012): 18–33.

Simmel, Georg. "Bridge and Door." Translated by Mark Rotter. *Theory, Culture and Society* 11 (1994): 5–10.

Simondon, Gilbert. "Technical Mentality." *Parrhesia* 7 (2009): 17–27.

Slezkine, Yuri. *The Jewish Century.* Princeton: Princeton University Press, 2004.

Sloterdijk, Peter. *In the World Interior of Capital:For a Philosophical Theory of Globalization*. Translated by Wieland Hoban. Cambridge: Polity, 2013.

Smeyers, Maurits. *Flemish Miniatures from the Eighth to the Mid-Sixteenth Century*. Leuven, Belgium: Davidsfonds, 1999.

Sophocles. *The Three Theban Plays: Antigone, Oedipus the King, Oedipus at Colonus*. Translated by Robert Fagles. Harmondsworth: Penguin, 1982.

Stevin, Simon. *The Haven-Finding Art, or, The Way to Find any Hauen or Place at Sea by the Latitude and Variation*. Translated from the Dutch by E. Wright. London: G.B.R.N. and R.B., 1599.

Stoichita, Victor. *The Self-Aware Image: An Insight into Early Modern Meta-Painting*. Cambridge: Cambridge University Press, 1997.

Szemerényi, Oswald. *Richtungen der modernen Sprachwissenschaft*. Vol. 1. Heidelberg: Winter, 1971.

Tambiah, S. J. "On Flying Witches and Flying Canoes: The Coding of Male and Female Values." In *The Kula: New Perspectives on Massim Exchange*, edited by Jerry W. Leach and Edmund Leach, 171–200. Cambridge, New York, and Melbourne: Cambridge University Press, 1983.

Thibaudet, Albert. *Gustave Flaubert, 1821–1880: Sa Vie—Ses Romans—Son Style*. Paris: Plon-Nourrit et Cie, 1922.

Thoss, Dagmar. "Georg Hoefnagel und seine Beziehungen zur Gent-Brügger Buchmalerei." *Jahrbuch der Kunsthistorischen Sammlungen in Wien* 82/83 (1986/87): 199–211.

Thrower, Norman J. *Original Survey and Land Subdivision: A Comparative Study of the Form and Effect of Contrasting Cadastral Surveys*. Chicago: Rand McNally, 1966.

Thürlemann, Felix. *Robert Campin, das Merode-Triptychon: Ein Hochzeitsbild für Peter Engelbrecht und Gretchen Schrinmechers aus Köln*. Frankfurt/M.: Fischer Taschenbuch Verlag, 1997.

Torpey, John. *The Invention of the Passport: Surveillance, Citizenship and the State*. Cambridge: Cambridge University Press, 2000.

United States Continental Congress. "An Ordinance for Ascertaining the Mode of Disposing of Lands in the Western Territory (20 May 1785)." In *Journals of the Continental Congress, 1774–1789*, vol. 28 (1785), edited from the original records in the Library of Congress by John C. Fitzpatrick, 375–381. Washington: U. S. Government Printing Office, 1933.

Vagt, Christina. *Geschickte Sprünge: Physik und Medium bei Martin Heidegger*. Zürich: Diaphanes, 2010.

Valéry, Paul. "Eupalinos, or, The Architect." In *The Collected Works of Paul Valéry*, vol. 4, edited by Jackson Matthews. New York: Pantheon, 1956.

Vasari, Giorgio. *Vasari on Technique: Being the Introduction to the Three Arts of Design, Architecture, Sculpture and Painting, Prefixed to the Lives of the Most Excellent Painters, Sculptors and Architects*. Translated by Louisa Maclehose and edited by G. Baldwin Brown. London: J. M. Dent, 1907.

Veitia Linage, Joseph de. *Norte de la Contratación de las Indias occidentales*. Seville: Blas, 1672.

Virilio, Paul. *Bunker archeology: Etude sur l'espace militaire européen de la Seconde Guerre mondiale*. Paris: CCI, 1975.

——.*Guerre et cinéma: Logistique de la perception*. Paris: Cahiers du cinéma, 1984.

——. *War and Cinema: The Logistics of Perception*. Translated by Patrick Camiller. London: Verso, 1989.

——. *Bunker Archeology*. Translated by George Collins. New York: Princeton Architectural Press, 1994.

Vismann, Cornelia. *Files: Law and Media Technology*. Translated by Geoffrey Winthrop-Young. Stanford: Stanford University Press, 2008.

——. "Kulturtechniken und Souveränität." *Zeitschrift für Medien- und Kulturforschung* no. 1 (2010), 171–82.

——. "Cultural Techniques and Sovereignty." *Theory, Culture and Society* 30, no. 6 (2013): 83–93.

Vogl, Joseph. "Becoming-media: Galileo's Telescope." *Grey Room* 29 (2007): 14–25.

Wasner, Franz. "Fifteenth-Century Texts on the Ceremonial of the Papal 'Legatus a latere.'" *Traditio: Studies in Ancient and Medieval History, Thought and Religion* 14 (1958): 295–358.

Wellbery, David. "Foreword." In *Discourse Networks 1800/1900*, by Friedrich A. Kittler, translated by Michael Metteer and Chris Cullens, vii–xxxiii. Stanford: Stanford University Press, 1990.

Westfehling, Uwe. *Zeichnen in der Renaissance: Entwicklung, Techniken, Formen, Themen*. Cologne: DuMont, 1993.

Wiemers, Michael. *Bildform und Werkgenese: Studien zur zeichnerischen Bildvorbereitung in der italienischen Malerei zwischen 1450 und 1490*. Kunstwissenschaftliche Studien, vol. 67. Munich and Berlin: Deutscher Kunstverlag, 1996.

Wigal, Donald. *Historic Maritime Maps Used for Historic Exploration, 1290–1699*. New York: Parkstone Press, 2000.

Wills, David. *Dorsality: Thinking Back through Technology and Politics*. Minneapolis: University of Minnesota Press, 2008.

Winthrop-Young, Geoffrey. "Going Postal to Deliver Subjects: Remarks on a German Postal Apriori." *Angelaki* 7, no. 3 (2002): 143–58.

——. "Mensch, Medien, Körper, Kehre: Zum posthumanistischen Immerschon." *Philosophische Rundschau* 56, no. 1 (2009): 1–16.

——. "Krautrock, Heidegger, Bogeyman: Kittler in the Anglosphere." *Thesis Eleven* 107, no. 1 (2011): 6–20.

——. "Cultural Techniques: Preliminary Remarks." *Theory, Culture and Society* 30, no. 6 (2013): 3–19.

Wolfe, Cary. *What Is Posthumanism?* Minneapolis: University of Minnesota Press, 2010.

——. *Before the Law: Humans and Other Animals in a Biopolitical Frame*. Chicago: University of Chicago Press, 2013.

Woodward, David. "Medieval Mappaemundi." In Harley and Woodward, eds., *Cartography in Prehistoric, Ancient, and Medieval Europe and the Mediterranean*, 286–370.

Wouk, Edward. "Dirk Bouts's Last Supper Altarpiece and the Sacrament van Mirakel at Louvain." *Immediations: The Research Journal of the Courtauld Institute of Art* 1, no. 2 (2005): 40–53.

Xenophon. *"Memorabilia" and "Oeconomicus."* Translated by E. C. Marchand. Cambridge: Harvard University Press, Loeb Classical Library, 1938.

Yasur-Landau, Assaf. "On Birds and Dragons: A Note on the Sea Peoples and Mycenaean Ships." In *Pax Hethitica: Studies on the Hittites and Their Neighbours in Honour of Itamar Singer*, edited by Yoram Cohen, Amir Gilan, and Jared L. Miller, 399–410. Wiesbaden: Harrassowitz, 2010.

Žižek, Slavoj. "'I Hear You with My Eyes,' or, The Invisible Master." In *Gaze and Voice as Love Objects*, edited by Renata Salecl and Slavoj Žižek, 90–126. Durham, N.C.: Duke University Press, 1996.

INDEX

MEANING SYSTEMS

The Beginning of Heaven and Earth Has No Name: Seven Days with Second-Order Cybernetics. Edited by Albert Müller and Karl H. Müller. Translated by Elinor Rooks and Michael Kasenbacher.
HEINZ VON FOERSTER

Cultural Techniques: Grids, Filters, Doors, and Other Articulations of the Real. Translated by Geoffrey Winthrop-Young.
BERNHARD SIEGERT